FINANCIAL EVALUATION OF ENVIRONMENTAL INVESTMENTS

FINANCIAL EVALUATION OF ENVIRONMENTAL INVESTMENTS

Tuula Moilanen and Christopher Martin

INSTITUTION OF CHEMICAL ENGINEERS

Published by
Institution of Chemical Engineers,
Davis Building,
165–189 Railway Terrace,
Rugby, Warwickshire CV21 3HQ, UK.
IChemE is a Registered Charity

© 1996 Tuula Moilanen and Christopher Martin

ISBN 0 85295 365 8

Printed in the United Kingdom by Galliard (Printers) Ltd, Great Yarmouth

To Risto John

FOREWORD

Much has been written about business and the environment — and rightly so. Environmental issues are now one of the key driving forces shaping the global economy and therefore are regularly encountered by most businesses and their managers.

Many business managers and environmentalists already argue that environmental best practice improves the financial performance of a company. This book is not making that claim but rather provides a tool to test the assertion. It thus provides a ground-breaking contribution to the debate. Perhaps more importantly, it provides a practical tool to assist managers in real decision-making situations. It does so by helping managers to find answers to one of the key questions about an environmental investment — what is the impact of the investment on the bottom-line of the company?

This book is not a theoretical discussion about environmental issues. It provides a practical management tool which takes some of the mythology out of environmental decision-making in companies, and also helps to assess the environmental questions in the traditional business context of financial imperatives and constraints.

By taking into account the long-term environmental implications of a decision and by translating this into the concepts of measuring financial performance, the authors provide a service to environmentalists and business alike. If — as I believe to be the case — a good environmental decision generally leads to better financial performance of a company, then the model and methodology presented in this book will guide managers towards better decisions. This will benefit both the company concerned and the physical environment in which we live.

Paul Gilding
Former Executive Director of Greenpeace International

SUMMARY

Increased public awareness of environmental pollution and its causes in recent years has meant that 'success' in the process industries is no longer simply a matter of business performance. Companies are under increasing pressure from environmental groups, governments, shareholders and consumers to improve environmental performance. But this improvement is not easy to quantify; it is determined in terms of compliance with legislation, cleaner process technologies, the formulation of products that are 'friendlier' to the environment, and promotion of the company's 'green' image.

Companies are faced with strategic choices when they try to respond to the requirements and demands of the environmental 'conscience'. Should they, for example, alter the process or change the product design ... or both? How can they incorporate environmental aspects such as possible future legislation, long-term liabilities and the expectations of various pressure groups into their decision-making process? The challenge is to build these environmental business 'drivers' into the company's business strategy so that a value added component is achieved for all stakeholders — consumers and shareholders alike.

Traditional investment analysis techniques such as 'net present value' or 'payback' calculations are widely used in the process industries. These are inadequate, however, when environmental issues have to be considered in the analysis. First, these techniques do not require that the project evaluator includes all the necessary and relevant factors. For example, considerations of possible future liabilities and the long time-scale needed to estimate the environmental impact of a product or process are not properly accommodated. Second, these techniques do not incorporate risk in a structured way that is both mathematically *and* logically correct. Abnormal discount rates or hurdle rates are sometimes used to compensate for this, but they do not provide satisfactory solutions.

Financial Evaluation of Environmental Investments addresses these problems. It takes a holistic view of decision-making — one that highlights the financial and economic impact of the environmental aspects of an investment. It provides a conceptual and practical methodology that allows the *expected monetary values* of different investment and risk scenarios to be encompassed in a mathematical model, a model that:

- includes wider sets of cost-benefit items than traditional techniques;
- deals with risk and uncertainty in a mathematically disciplined way. It guides and helps the user to incorporate these into each item to be included in the appraisal;
- considers specifically the environmental aspects of every facet of an investment (particularly the long time-scale required to determine any environmental impact);
- helps to quantify less tangible items such as the impact of the investment on corporate image, financing cost and future liabilities;
- incorporates into the analysis as direct investment-related costs items such as pollution monitoring that are traditionally considered as overheads.

Financial Evaluation of Environmental Investments is a practical, user's guide to environmental investment analysis in the process industries. Chapter 1 deals with the context and introduces the basic concepts. Development of a conceptual model and a methodology is discussed in Chapter 2. A detailed description of the various modules in the model is given in Chapter 3. Two test cases are presented in Chapter 4 to illustrate implementation of the methodology in specific situations. Appendix 1 contains the check-lists used to conduct the actual analysis. The expected monetary value technique and its application are discussed in detail in Appendix 2.

BACKGROUND TO THE BOOK

Financial Evaluation of Environmental Investments is based on work carried out by Tuula Moilanen and Christopher Martin, on a project commissioned by the consortium of the Intelligent Manufacturing Systems (IMS) Test Case 2. Their task was to find ways of evaluating the financial consequences of clean technologies. The IMS project is a global ten-year programme for pre-competitive collaborative R&D projects, and was initiated by the Ministry of International Trade and Industry in Japan. The Test Case 2 'Clean Manufacturing in the Process Industries' consortium consisted of ICI, DuPont, Teijin, VTT, Foster Wheeler, John Brown and the Finnish Forest Industry Federation.

In addition to the book, Tuula Moilanen has written an IChemE training package on the same theme, entitled Environmental Investment Appraisal. For further information on the package, please contact IChemE, Davis Building, 165–189 Railway Terrace, Rugby, Warwicks CV21 3HQ, UK (tel: 01788 578214, fax: 01788 560833).

ACKNOWLEDGEMENTS

The authors wish to thank the following for their valuable contribution to this study:

- Chalmers University of Technology in Gothenburg, Sweden, and particularly Ms Henrikke Bauman and Ms Anne-Marie Tillman;
- The German Federal Environment Agency, and particularly Mr Ruckerstein in the Environmental Economics Department;
- The Tellus Institute, USA, and particularly Mr Allen White;
- The US Environmental Protection Agency/Pollution Prevention Research Branch, and particularly Mr Harry Freeman;
- The US Embassy in London, and particularly Mr Nigel Standen;
- The Embassy of Canada in London;
- The Embassy of Finland in Tokyo, and particularly Mr Esa Moilanen;
- Kymmene Corporation and Wisaforest Oy Ab, and particularly Mr Kari Ebeling and Mr Paavo Heikkilä;
- Foster Wheeler Energy Ltd, and particularly Mr Andy Allen;
- ICI Katalco, and particularly Dr Martin Fakley;
- Mr John Swarbrick, a former colleague, who assisted in the preparation of Test Case B.

CONTENTS

Table of contents

Topic	Description, section/page	Check-list/ page	Test case/ page	Appendix/ page
R&D	2.2/10	A/107	A/72	
Product design	2.3/18	B/121		
Production	2.4/23	C/131	B/95	
Marketing	2.5/32	D/148	A/75	
Company image	2.6/35	E/159	B/95	
Expected monetary value		102	A/76	2/163
Discounting	2.7.2/45	101		
Traditional assessment	3/51			
Overheads	2.7.4/47	106	A/74	

1.　INTRODUCTION

Process industries in the Western industrialized world are under continuous pressure to improve their environmental performance. In the UK, for instance, the membership of environmental groups increased from two to four million from 1980 to 1990. Environmental performance may be measured in different ways — compliance to legislation, level of discharges from production processes, the environmental image of products and so on. The pressure is, in turn, exercised by various environmental groups, governments, shareholders and consumers.

Increasing expectations to improve environmental performance means that companies not only have to find ways to clean the emissions from existing processes, but also need to respond to the requirements and demands of various interest groups. Therefore companies in the process industries must choose from various options such as altering the production process and/or changing the product design, so that the production process is cleaner and the end-product causes minimal harm to the environment.

When faced with such choices a company needs a decision-support tool that offers the opportunity to evaluate the risks, returns and costs associated with the various options. These choices are strategic in nature and require an overall evaluation of the competitive situation and the expected regulatory environment. Consideration must also be paid to the expectations of various interest groups.

This book aims to provide a conceptual model and a practical methodology based on the model that can be developed into a mathematical representation of the expected monetary values (EMVs) of different investment and risk scenarios. The model encompasses many of the aspects to be considered when an investment decision is to be made. The investment may be in:
- research and development;
- product or process design;
- new process technology;
- promotion of the 'green' image of a product or a company.

In each case a company needs to understand the costs, the returns and the uncertainties of the various parameters in the investment and external environment that affect the outcome of the investment. The model aims to highlight the financial and economic impact of environmental aspects of the investment.

Some traditional techniques are available for evaluating the financial impact of investments. These techniques, such as net present value (NPV) or payback calculations, are widely used in the process industries despite their inadequacies. These inadequacies are highlighted when the environmental aspect of an investment is incorporated in the analysis.

The traditional techniques do not easily allow for the inclusion of risk in the analysis. This has led to abnormal discount rates or hurdle rates being used. Existing models do not direct evaluators to incorporate all the necessary and relevant factors influencing the financial outcome of the investment under study. Nor do they assist in approaching risk and uncertainty in a structured fashion. Investments with environmental aspects suffer unduly if traditional methods are used. This is because the unusual but necessary considerations of future liabilities and the long time horizon of an environmental impact to materialize are not properly incorporated in traditional analysis.

The model presented here and the methodology based on it aim to:
- be *easily used* by anyone accustomed to appraising capital projects;
- be *generally applicable* to any process industry investment;
- use *monetary units* as common accounting units;
- include a mechanism to account for *risks and uncertainties*;
- form a basis for a database of *technology costs and benefits*, be they environmental or not;
- assist in incorporating *environmental aspects and their financial impact* into the analysis.

The system being studied is a company encompassing its assets, its employees and their actions. Broadly speaking, the environment is everything else outside this system (see Figure 1.1). For the purposes of the model, environment is defined as the natural, biological environment. The rest of the items outside the company are left uncategorized. The fundamental assumptions of the model and the use of the methodology are:

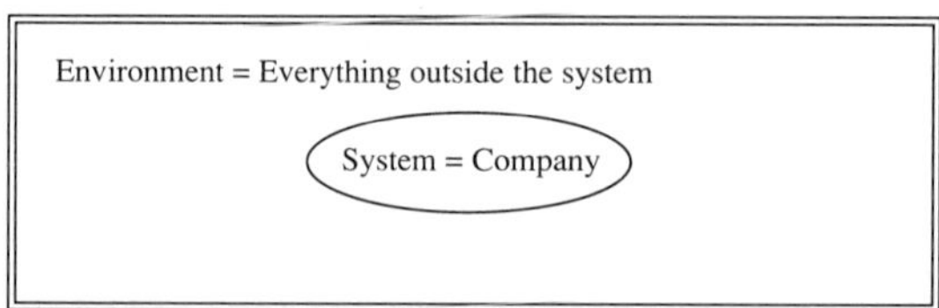

Figure 1.1   The system and its environment.

(1) Only future cash flows are incorporated and each parameter is assigned a risk factor. This approach allows and ensures that only the relevant parameters are included in the calculations. Items such as sunk costs are, therefore, excluded. It also ensures that all parameters are compared and brought into the equation on a similar basis. The fact that each parameter is allowed to have its own probability distribution means that no artificially high discount rates are required.

(2) Externalities — that is, parameters outside the company — are not included if they do not have a cash flow impact on the company. A pollution tax is an example. Externalities are defined loosely and only to highlight which parameters are inside the company and which are outside. In many cases, there needs to be a mechanism in place for importing the financial value of an externality into the company's investment equation. These issues will be discussed later.

(3) The full impact on all parts of the company, such as on company image, is considered along with the technical aspects of the investment. It is natural that project evaluators concentrate on the issues that they are most familiar with. The idea of this model is draw the project evaluator's attention to all the other issues including the less familiar ones.

(4) The analysis should be conducted in co-operation with various departments and parts of a company. Traditionally, the department of the company preparing the capital allowance application or investment plan has worked in isolation and with relatively little consultation with other parts of the company. As investments are often large and complex, particularly in the process industries, it can be beneficial to call upon all the inside and outside expertise available.

(5) Life cycle analysis (LCA) and other environmental management tools 'feed' the model with information on the environmental impact of an investment. Many of the parameters that need to be taken into account in the investment decision require extensive studies. LCAs may take years to complete and provide extensive numerical information. This information needs to be expressed in terms of financial impact, which can be a complicated task. It is therefore natural that these calculations take place outside the model.

1.1 INVESTMENTS UNDER STUDY

Investments that can be analysed using this model are:
- those which may have an impact on the environment and the environmental performance of a company;
- those which are made specifically to achieve an environmental goal.

The investments are defined broadly to cover both capital investments and investments in activities that can be in the area of:

- R&D projects;
- product design projects;
- process technology;
- marketability and marketing;
- company image and corporate communications.

The challenge is to identify the ultimate purpose of an investment. For example, a product design change may be required to enhance sales, or process changes may be required to improve the environmental image of a company. Each investment has far-reaching consequences, but these may be ignored or bypassed if the full scale of the investment purpose is not clearly defined.

The methodology assists in identifying the financial impact of the environmental aspects of an investment, independent of the ultimate goal and purpose of that investment. Environmental aspects can be, for example, impact on:

- effluent streams;
- recyclability of process by-products;
- environmental qualities of the end-product;
- the environmental image of the company;
- the environmental risks of the company.

In the case of investments that are not specifically targeted at achieving an environmental goal, the challenge for a company is twofold. First, the environmental impact must be identified as a separate task from the overall investment evaluation. Second, the cost of an investment in the impact, and the respective return, need to be valued in monetary terms. These challenges are also present in environmentally-focused investments. However, the initial recognition of the need to value the environmental impact is already explicit.

The cost of an investment is measured from the point of view of a company, not society. The cost is therefore defined as the out-of-pocket expenses that accrue to the company making the investment. Externalities, such as the societal cost of 'lost' clean air, are not taken into the equation initially unless there is a tax or fee for polluting the air — which a company has to pay — or some other quantifiable negative or positive impact on future revenue.

Return on investment in an environmental aspect of a product or production may be difficult to identify and requires special attention. Identification of these returns either tends to be neglected in analyses currently conducted by companies, or is only partially considered due to perceived uncertainties. An important approach to the identification of returns is the comparison of various investment alternatives and their costs and returns in general, and those of environmental aspects of an investment in particular.

4

1.2   CALCULATION BASIS

Investment analysis can be done either on an incremental basis or by comparing costs resulting from different alternatives. When an analysis is carried out an incremental basis, only the change in costs and revenues — or totally new cash flows — are taken into account. This approach is especially useful if the investment under study is an improvement to an existing unit operation or product design.

Where the same goal can be achieved through different investment alternatives — such as building a totally new plant or just adding an end-of-pipe treatment to an existing plant — it is advisable to compare the cost and return implications in isolation. In this case the cost impact on either overall product cost or operating costs should be compared for two or more alternatives. If the cost base changes, it can be expected that the revenue streams will change as well.

In most companies investments are mutually exclusive, as the funds available for investment are limited. This means that the opportunities gained and lost by adopting a particular investment route must be considered. For instance, an investment in a clean process technology may mean that taxes, fees and waste treatment costs are reduced or avoided. An investment not executed may mean that potential sales are lost.

1.3   TECHNICAL BASIS OF THE METHODOLOGY

The methodology is based on present knowledge of dynamic investment appraisal methods, and the traditional approach is broadened and complemented. The model offers a holistic picture of all the aspects contributing to the outcome of the investment by evaluating a process decision not only from a technological point of view but also in the broader context of the overall competitive position. Thus, a larger-than-traditional set of costs and returns is taken into account.

The major problem in appraising an investment in mathematical and, eventually, monetary terms arises from the uncertainties in which a company operates. These uncertainties need to be expressed in terms of a risk measure which is, by its nature, subjective. The suggested technique is the expected monetary value (EMV) calculation. This technique requires identification of the probabilities of each possible outcome of each parameter. Although potentially cumbersome, the technique is very powerful and has been used in various industries for estimating the potential outcome of risk situations, such as oil exploration and protecting cargo vessels from enemy submarines during the Second World War.

1.4   WHY IS THIS MODEL AND METHODOLOGY DIFFERENT?

Engineers and accountants have been trained in traditional investment analysis and have applied the traditional techniques with varying degrees of success. Both these professional groups are familiar with the basic decision rules of the traditional techniques. Why is a new approach required?

This model and its practical application methodology add to the traditional techniques, because:

(1) The *sets of cost and benefit* items included in the analysis are broader than in traditional techniques. The model's construction and mode of use assists in finding new cost and benefit categories.

(2) *Risk and uncertainty* is dealt with in a mathematically disciplined fashion and assistance is given in structuring uncertainty around each individual cost and return item included in the appraisal.

(3) The specific *nature of environmental aspects of investments* is taken into account in every part of the investment under study. This refers particularly to the long time-scale required for some environmental aspects to materialize.

(4) The methodology assists in *quantifying items that are usually left unquantified*. These are items such as the impact of the investment on corporate image, financing cost and future liabilities.

(5) Traditional *overhead items* are incorporated into the investment analysis as direct, investment-related costs.

(6) Use of the model ensures that a holistic, company-wide picture of the investment is gained. The methodology ensures further that the various feedback loops and interdependencies between investments and activities are taken into account. This can be seen in Figure 1.2.

1.5   DEFINITIONS

1.5.1   INVESTMENTS

Investments are defined in broad terms to cover both traditional capital investments and investments in activities, such as marketing or research and development. Although the financial accounting rules for treating capital and other investments are different, from an appraisal point of view there is little difference. The only difference arises in the taxation treatment, which can be significant, but this should be evaluated in collaboration with the company

6

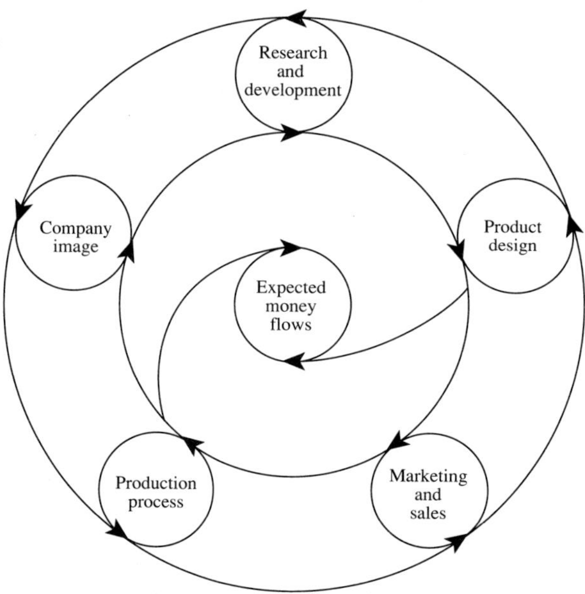

Figure 1.2   Feedback loops and interdependencies.

accountants. Some capital investments may be eligible for subsidized financing, which will be a revenue item to be considered.

1.5.2   COSTS

The concept of cost is used interchangeably with words such as expenditure, investment and out-of-pocket expenses. The exact choice of words depends on the accounting context and the point of view taken. For instance, any investment is an expense in an accounting context. Whether the expense is recognized in full in the profit and loss account in the first year, or in future years through depreciation charges, depends on the expectation of the timing of the return from the investment. In this context, the accounting differences can be ignored. Taxes are not considered either.

1.5.3   RETURN

The concept of return is used interchangeably with words such as revenue, income, benefit and profit. Again, in accounting terms these concepts have various interpretations. What is important in this context is the positive effect a parameter has on the company in monetary terms. The effect can be measured at different levels of a profit and loss account, but nonetheless it is positive.

1.5.4   RISK

The concept of risk is used interchangeably with the word uncertainty. Although in some cases these words are not synonymous, for the purposes of this methodology they are. Uncertainty means risks for a company, and in a way risk measures uncertainty. The types of risk considered in this study include normal business risks that can negatively affect a company. They do not include risks that may force a company to cease operations. A company tries to find those investments, among many options, that bring the most positive return with the least negative impact. A company does not knowingly engage in activities it knows have an unacceptable probability of threatening its very existence (see Figure 1.3).

1.5.5   EXTERNALITIES

Externalities in a traditional microeconomic sense are spill-over effects, third-party effects, neighbourhood effects or, simply, external effects. Essentially they are involuntary by nature. The fact that a gardener waters a lawn in a strong wind and therefore water also ends up on the neighbour's lawn is most likely involuntary and external, outside the lawn that the gardener wants to water in the first place. Externalities are costs and benefits that do not enter fully and appropriately into a decision-maker's equation. Often there needs to be a mechanism to import the financial impact of these externalities, particularly if they relate to environmental impacts, into an investment equation.

1.5.6   ENVIRONMENT

Environment can be defined in many ways. In this book environment means the natural and biological sphere of our globe where human, animal and plant life can continue, or which contributes to the possibility of human, animal and plant life continuing. So here environment comprises water, air and soil but also the

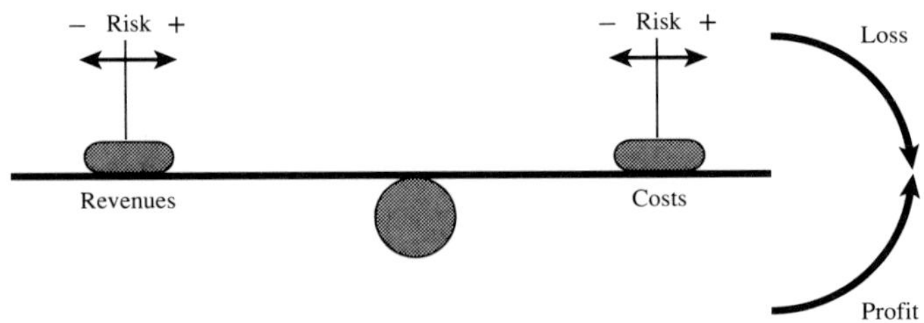

Figure 1.3   Finding the balance.

social and cultural aspects of human life. Pollution and contamination are used interchangeably, although in certain cases they are not interchangeable.

1.6 HOW TO USE THIS BOOK

This book aims to introduce the relevant concepts and the structure of the model and the methodology based on it. The objective is that the reader should actually start using the model and the methodology in real-life situations as soon as possible. Therefore two real-life examples are described with accompanying calculations.

Appendix 1 contains a user guide with the relevant check-lists for each module in the model. The pages can be used time and again for different projects. It is also possible to purchase the training package based on this book, entitled Environmental Investment Appraisal, from the Institution of Chemical Engineers.

Appendix 2 explains the EMV technique. A software package containing the required program is included in the training package. But there is no computer model or software package available to perform all the calculations. It is suggested that a worksheet program, such as Excel or Lotus, is used for compiling all the information. There is no particular list of sources for technology cost or LCA data. These vary greatly from industry to industry, country to country and product to product. Thus an exhaustive list of databanks and software would be impossible to produce.

2. DESCRIPTION OF THE MODEL AND THE METHODOLOGY

2.1 GENERAL CONSTRUCTION

The model and the methodology comprise several modules that describe the areas where an environmental investment has an impact. Take an investment in a product design as an example. First, the product is developed based on some research. Then the product is designed, bearing in mind the R&D results and production requirements. In the next stage the product is actually produced and, finally, it is sold.

At each stage a company needs to take account of the expectations of its stakeholders, be they company employees, shareholders, the general public or the regulator. The actions of a company influence the perception the stakeholders have of the environmental performance of a company, thus forming a part of the company image.

The structure of the model and methodology and the interaction between various modules are shown in Figure 2.1. This illustrates some of the interdependencies between the modules and the various factors influencing each module. The methodology can be used in different decision-making situations, whether it is a question of choosing one among several R&D projects, selecting a new process technology, defining a product portfolio or establishing a new marketing strategy. Each of these 'investments' has costs and returns that arise in different parts of a company. For each area of responsibility there is a module that aims to detail the costs, returns and risks to be included.

Some companies use an asset or a plant as the basis for strategic and operative decision-making. All decisions and performance measures are based on the entity of a plant. That does not exclude these companies from using the model. It may mean that they will place more emphasis on the production module of the model, but equally the companies need to consider the impact of their actions on the environment and the company image, for instance.

2.2 RESEARCH AND DEVELOPMENT

Investments in R&D may have a number of goals, including:

- pollution prevention;
- cost reduction;
- increasing profit.

These goals may be achieved by developing new products, processes or saleable technology. Both general R&D projects and R&D projects which are specifically aimed at improving the environmental performance of a company are considered. This underlines the importance of evaluation of the environmental impacts of all R&D projects. R&D may contribute to an investment with or without a specific environmental goal. The R&D under study may constitute the actual investment or only a part of a larger investment. In many cases investment in R&D can be regarded as a sunk cost. This is the case when the R&D effort has already taken place before the investment under study is to be undertaken. Sunk costs should not be taken into account in the investment appraisal.

The emphasis in this approach is on development rather than research. Research is here defined as the process of invention which can produce a product or a process. By its nature invention is less targeted than development work. Therefore targeted development work rather than research for new innovations is considered in investment appraisals.

Evaluation of R&D projects tends to depend on the main object of the project. This has been established in questionnaire-based research conducted in

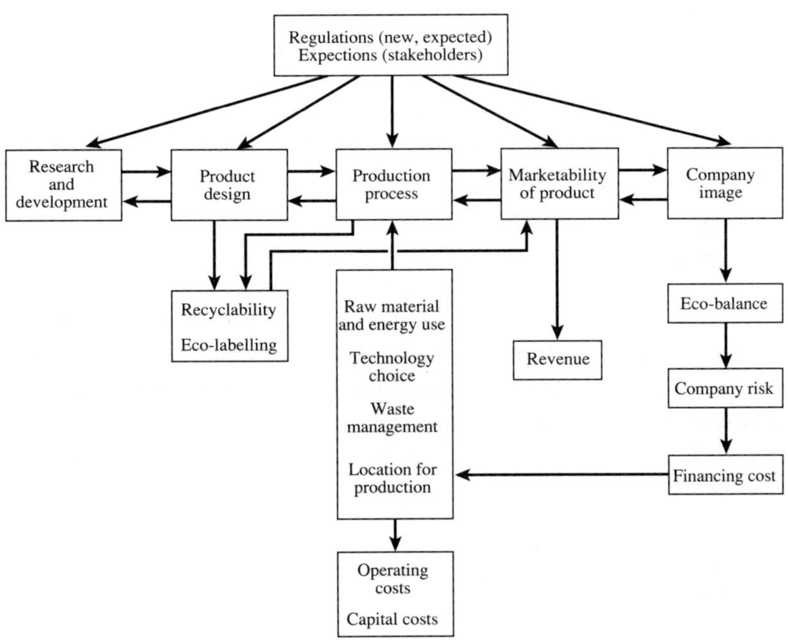

Figure 2.1   Structure of methodology.

Germany and the USA[1,2]. Projects that have cost reduction or profit sustenance as their main target have not generally been evaluated for their specific environmental impact. Objectives of environmentally-focused R&D can include:

- reduction of emissions from a product and its production process;
- reduction in the use of energy and raw materials;
- increase in process internal recyclability of resources;
- increase in recyclability of the end-product;
- reduction of the environmental impact of the end-product;
- reduction in health risks associated with the product and its production process;
- longer product lifetime in use.

All these aspects have a different impact on the producing company. These aspects may:

- allow cost reduction;
- increase productivity;
- fuel new product and process innovations that give technology leadership and hence a better market position;
- improve the environmental image of the company and its products;
- lower the overall risk in operations so that liability costs are likely to be lower with environmentally-friendlier products and processes.

Each of these aspects improves the overall performance of a company in its own way.

If R&D is targeted at producing a new technology which will alter the competitive position of a product through superior utility or lower cost, this R&D is likely to be done in-house. This is particularly true if the technology is patentable. If pollution prevention is to meet some particular corporate or regulatory objectives, in-house R&D is only one way of achieving the necessary level of process or product change. It is also possible to buy in the necessary technology. This may mean purchasing patents or licences. In-house R&D can also be directed either to changing the process or product design, or to investing in waste treatment at the end of the process. In cases where a company needs to choose between a new, clean technology or keeping the old one and adding a waste treatment facility, an R&D project may need to be compared with various alternative approaches to solving the problem. A summary of the R&D module is given in Figure 2.2 and a short example in Box 1 (see page 14).

2.2.1   COSTS OF ENVIRONMENTAL ASPECTS OF R&D

ENVIRONMENTALLY-FOCUSED R&D

In environmentally-focused R&D, the cost of influencing the environment is easily defined: it is the total cost of the project in question. The traditional cost

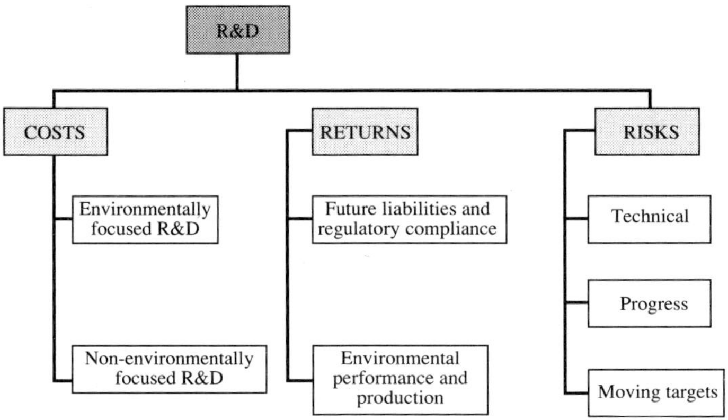

Figure 2.2   The R&D module.

elements are direct labour, equipment and testing. A difficulty arises in estimating the overall time required for the project. Another difficult area, and therefore often excluded from traditional analysis, is testing costs. Many companies do not include these in calculations, although they may represent a major expense. They should be incorporated in the analysis methodology described here.

The total cost of an R&D project is not only the time and equipment used for the project but also the necessary investments required for implementing the results of the project. Production based on the results of R&D will provide a company with either positive or negative returns. These returns depend on factors such as marketing, product quality and market conditions. This model incorporates these costs in the other modules.

NON-ENVIRONMENTALLY-FOCUSED R&D

When the main objective of an R&D investment is to reduce production costs or sustain profitability, it may be difficult to identify the environmental portion of the R&D investment. It may be argued that it is unnecessary to separate these cost items from the others, but clearly the full benefits of an R&D investment are not accounted for if the respective environmental benefits are not considered. A company may receive positive financial returns in the form of increased sales, for instance.

Even if a company has no environmental policy or specific environmental targets, it is possible to evaluate the environmental impact of the results of an R&D project. The project changes either a production process or a product,

> ### BOX 1:
> ### What a Waterlily!
>
> *ICI Polyurethanes was producing rigid foams used mainly in the car industry. The company wanted to enter a new market for flexible foams but was faced with the problem that CFCs were being used in the foaming process. ICI had to find a replacement. A new product, 'Waterlily', was developed. Not only did it provide an alternative to CFCs but also allowed the foam to be recycled.*
>
> *How did ICI go about finding the solution to an essentially marketing problem? With the help of an LCA, ICI was able to establish the environmental problems with the existing flexible foams. Although CFCs were present in the product composition, it was the waste created by the number of discarded mattresses and other products for which foams were used that led ICI to focus its R&D efforts.*
>
> *With the help of a German company, Metzeler Schaum GmbH, ICI started studying how a future recycling scheme would operate. In order to optimize the recovery of polyurethane, the scope of the study was limited to production of scrap and post-consumption waste from 'full foam' mattresses in the institutional market. Particularly, the hospital market was studied as the mattresses conform to a well-defined standard, and the mattresses are regularly washed and sterilized.*
>
> *ICI calculated the cost of collecting the mattresses and transporting them to recycling. The recycling loop had to be defined carefully and controlled closely to ensure that the right kind of polyurethane entered the recycling process. Additionally, easily identifiable chemical identification of the product and detection system had to be evaluated. These were complemented by a physical label.*
>
> *The concept of recyclable polyurethane has been successful not only because of the inventive chemical composition but because of the simultaneous development in recycling well-defined waste categories.*

and the changed process or product will have a different environmental impact from the former process or product. The methodology used by a company to evaluate the potential sources of environmental impact can be, for instance, life cycle analysis (LCA).

LCA provides a picture of the raw material and energy use, emission levels and waste management costs. It helps to reveal to what extent the environmental regulations are met or exceeded. This in turn gives an idea of the 'friendliness' of the product or process from an environmental point of view. These facts can be used in marketing and in evaluating the overall risks of a company — for instance, when evaluating the risk of liability law suits. The marketing and marketability questions are discussed in Section 2.5, page 32.

If an analysis of environmental impacts is not included, other than in the analysis of specific pollution prevention projects, many cost, risk and return items are omitted from the project appraisal. In practice, pollution prevention investments suffer unnecessarily when not all the costs and returns of all investment types are reviewed and highlighted. This has been established in studies conducted in the USA[3].

2.2.2   RETURNS OF ENVIRONMENTAL ASPECTS OF R&D

Returns from environmental aspects of R&D investments can be classified into five categories:
- achieving regulatory compliance (and thus continued operation and elimination of fines);
- improved environmental performance and thus improved sales and lower company risk;
- reduced costs or profit sustenance;
- new income sources from patents and licensing of new technology;
- grants and subsidized financing.

Most American companies interviewed in a study conducted by the Tellus Institute[4] gave regulatory compliance as the main reason for pollution prevention investments. The aim was to avoid future liability costs, such as fines, law suits, increased worker health costs and other long-term risks that may be caused by potential hazard to the environment. Evaluation of future environmental liabilities is essential if the full costs and returns of an R&D investment are to be exposed; these costs and returns can be significant.

FUTURE ENVIRONMENTAL LIABILITIES AND REGULATORY COMPLIANCE

The main problem in estimating these costs and returns is the degree to which the regulatory environment changes, particularly because these changes seem irregular. In many cases the history of the adoption of certain regulations in different countries is so short that it may be difficult to see any development pattern. Many larger companies have therefore set their own targets that often exceed the present regulatory requirements. The difficulty, then, is that the regulatory authorities may speed up the development in reducing emission levels. (See Test Case B on page 88 — this was one of the concerns of the ICI project team.)

An additional problem in estimating future regulatory requirements is that in many countries it is not clear how the regulations will be formulated, and what the measuring device is going to be. This is a particular problem in cases of very long-term investments: it is difficult to see the benefits of an investment when the goals are unclear. Depending on the situation, a company may be far from compliance or contemplating exceeding present requirements. In the former case a company knows what the cost of non-compliance is. In the latter case a company needs to develop its own estimate of the likely developments in the regulatory environment, and of the benefits of exceeding regulatory requirements. Indeed, there is a complex interplay between companies and regulators in establishing the regulatory environment.

Other future liability costs may arise from unforeseen hazards, either in the form of unwanted leakages of hazardous substances or bodily harm caused to employees. Insurance mathematics may assist in these cases. Insurances are usually taken to protect a company against these types of risk. The insurance cost of existing production and product liability can be used as a benchmark for evaluating the benefits of improving the safety and environmental performance of a process or end-product. The reduction in insurance cost is a direct benefit from an R&D project.

The full financial impact of a future liability may be difficult to estimate and often a surrogate measure must be found. This can be the cost of land remediation or an estimate of fines accrued. These estimates are made uncertain by the fact that there are only a few legal cases where actual fines have been levied. Discounting techniques come to the rescue here. The further away a future cash flow is, the less impact it will have on the present value of an investment. Thus only very large sums will be really significant. If a future liability is deemed to be so large that it will threaten the company's existence, the company should not engage in such an activity in the first place.

Regulatory compliance is a way in which companies can improve environmental performance, but other means are also available. For example, a product can be less environmentally hazardous, a production process may use less energy or internal substances could be recycled. These changes may improve the public perception of the environmental image of a company. Environmental image can be used as a marketing tool in those cases where it has a direct appeal to the end-user. As and when reporting on environmental performance is made more standardized and transparent, the stakeholders will be able to assess the environmental risk of the company to a much greater extent than they can today.

ENVIRONMENTAL PERFORMANCE AND PRODUCTION

Many process industry products are intermediates that are used in producing the actual end-product for a consumer. Even in these cases aspects such as recyclability can be crucial for the marketability of a product. For example, some substances — such as phosphates — may be totally forbidden in detergents, hence restricting the sales of the product. Company image affects the perception of risk held by insurers and other stakeholders. A good environmental performance, as measured either by perception or by actual eco-balances, may also have a direct cost impact on a company in the form of lower financing cost or insurance premiums. These factors are discussed in Section 2.6, page 35.

Product and process improvements resulting in more environmental friendliness may not only affect the operating cost of a process but also the overall production cost. These savings are direct benefits of R&D work done on changing the product or process. Although these savings are conceptually clear, there are many technical measurement problems. For instance, it may be difficult to measure the financial gain of being able to recycle a process substance, or of being able to sell a by-product if the by-product has not got a clear market value. Process cost evaluations are discussed in Section 2.4, page 23.

An R&D project may be conducted in order to lower the production costs in a process or to sustain a certain profit level. In these cases the R&D project may have some environmentally positive effects as a 'by-product'. In order to be able to identify these effects an LCA may be conducted to list all the emission levels and uses of energy and raw materials, and to estimate waste handling costs and so on. The analysis should be conducted for both the present process and the changed one in order to find the areas of saving. These are direct returns on the project.

2.2.3 EVALUATING RISKS IN ENVIRONMENTAL ASPECTS OF R&D

Generally speaking, risks associated with R&D projects include:
- technical risks;
- project progress risks;
- target-oriented risks.

Technical problems can arise if the required technical specifications cannot be achieved. Progress problems may depend on unforeseen technical problems, management difficulties or manning problems. Target problems occur when either external or internal factors cause the goal posts to be moved. Typically, environmental targets change rapidly.

Company internal changes to environmental targets may depend on the management's changing policies, shareholder pressure, employee demands on safety and health standards and relative relationships between cost categories. Although company policies are stable, at least in the medium term, change in management or in the external environment may cause changes in company policy. Shareholders may also exert pressure in order to lower their perception of the risk level associated with potential environmental problems. Employee groups may demand changes to safety and health standards due to, for example, some new scientific discoveries. The relative costs of raw materials, energy and capital versus labour may alter due to efficiency gains and losses.

External changes happen in the regulatory framework from the efforts of various pressure groups, through changes in consumer tastes and requirements, in trade agreements and negotiations, and in the economic environment. Developments in consumer tastes and expectations are a target for continuous market studies and should be conducted with the actual customers of the process industry. They will be able to give valuable information about the expectations of the end consumers. The outcome of negotiations on trade agreements is difficult to predict and therefore it is important that a company has a good knowledge of the costs of having to change from one market to another. The economic environment changes due to different rates of economic development in different countries.

All these factors are difficult to predict. *It is essential that the R&D department or function does not attempt to make these predications without consulting specialists inside or outside the company.* It has been established in Germany[5] that R&D staff have only limited knowledge of expected changes in the regulatory environment and derive this knowledge from non-expert sources. The R&D function cannot be expected to have knowledge of consumer tastes, trade negotiations between countries and expected changes in the economic environment in either bilateral or multilateral relationships. *The necessity of co-operation between various departments and functional areas in a company when a financial analysis of R&D projects is being made is clear.*

2.3 PRODUCT DESIGN

In this book product design is defined as consisting of product characteristics such as:

- physical composition;
- recyclability;
- performance in intended use;
- aesthetic design (if intended for end-consumer use);
- consumer perception of product characteristics.

> *BOX 2:*
> *Life Cycle Design of Environmentally-Friendly Products*
>
> *The US Environmental Protection Agency (EPA) has commis-*
> *sioned the development of a* Life Cycle Design Guidance Man-
> ual *(EPA/600/R–92/226 Office of Research and Development,*
> *Washington DC 20460, USA). The purpose of the manual is 'to*
> *facilitate the development of technologies and products that*
> *result in reduced aggregate generation of pollutants across all*
> *media'. According to the manual, life cycle design is defined*
> *as 'a proactive approach for integrating pollution prevention*
> *and resource conservation strategies into the development of*
> *more ecologically and economically sustainable product sys-*
> *tems'.*
>
> *Although it can be argued that the entire life cycle of a product*
> *is not a concern of the producer, product liability issues have*
> *brought this consideration into the overall investment equation*
> *of a company. In addition, some consumer segments are willing*
> *to consider issues such as recycling or to purchase goods that*
> *are environmentally friendlier both to them and in the produc-*
> *tion. An example of this is the success of The Bodyshop.*
>
> *The flow chart describing the life cycle highlights the fact that*
> *continuous feedback is needed between various parts of a com-*
> *pany from R&D to marketing (see Figure 2.3 on page 20). A*
> *company should not forget the expectations of consumers either.*

The environmental investment in product design is the effort put into changing one or several of these characteristics so that the product is perceived to be more environmentally friendly by consumers, regulators, other outside interest groups and the producing company. A company may wish to alter the product design to improve the product's environmental performance (see Box 2). A particular design change may be necessary to secure the viability of the product. A product may have to be designed due to environmental regulations, such as a replacement for CFCs.

Figure 2.4 on page 21 summarizes product design considerations.

2.3.1   COST OF ENVIRONMENTAL ASPECTS IN PRODUCT DESIGN
Changes in product design incur costs due to, for example:
- R&D required;
- designer effort needed;

- process changes caused;
- administrative costs from changes in purchase and supply contracts.

It is important to note that the total cost of a design change cannot be estimated within one function or department only, as the design change involves many parts of a company organization. Identification of the cost of R&D is discussed in Section 2.2, page 10. Designer effort can be either an in-house cost that has to be clearly identified through an internal order placement process, or an external cost for which a contractual cost can be easily identified.

Much effort has been invested in developing the concept and approach to life cycle design of products and product systems. This work is particularly

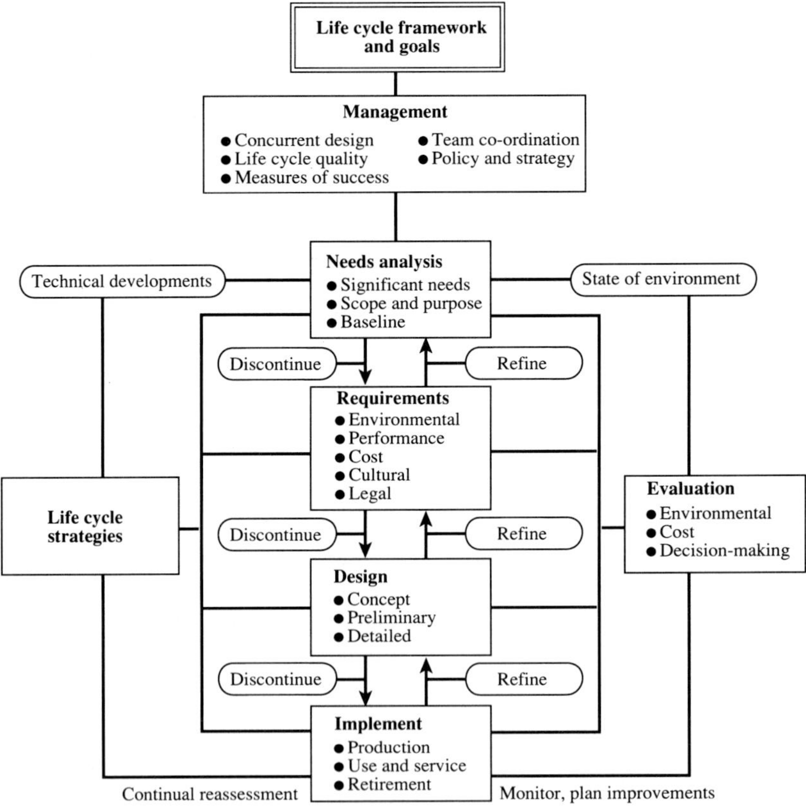

Figure 2.3   Life cycle design process. (Source: *EPA600/R–92/226 Life Cycle Design Guidance Manual*, January 1993.)

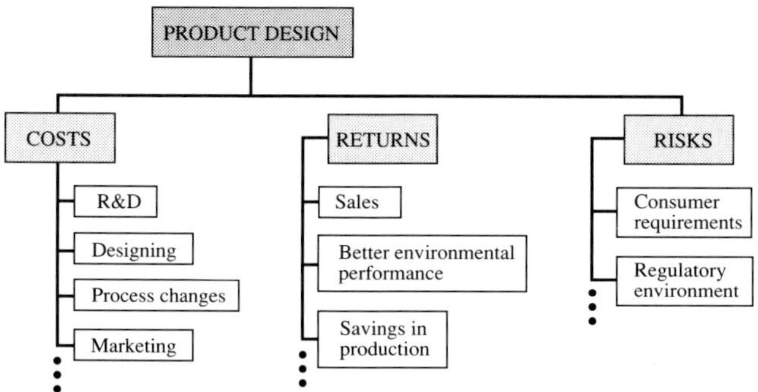

Figure 2.4   The product design module.

advanced in the US where the Environmental Protection Agency has financed the development of a manual for life cycle design (see Box 2 on page 19). This approach further highlights the need to work in cross-disciplinary teams and for concurrent design and total cost assessment. Central to this approach is the idea that the total impact of a product from cradle to grave is kept in mind and assessed as the product is being designed.

The cost of process changes required due to changes in product design may be difficult to estimate. The changes may involve:
• use of environmentally 'friendlier' and less energy-intensive raw materials and process materials, such as solvents and catalysts;
• substitution of non-renewable resources with renewable ones;
• minimization of waste;
• recycling of production materials such as cooling water or solvents;
• reduction of noise and odour from production.

All these aspects contribute to a new production cost equation that needs to be estimated on the basis of the technical description of the production process, such as flow sheets, material and energy balances, eco-balances and a technical description of the physical properties of the product. These can be used in conducting a LCA and in monetarization of the various emission levels, raw material uses and so on. LCA is discussed in Section 3.3, page 53.

A new product design may mean that the marketing concept for the product needs to be renewed. This may involve not only changes in advertising but also in marketing channels and markets covered. A new pricing policy may be needed. These issues must be addressed by the marketing department. In all, a totally new investment analysis that covers the whole life cycle of the newly-designed product needs to be conducted. This can be avoided if the change is very minor and affects only one particular area in the product's progress through a company.

A new design or a new product may also require a new service set-up. A fairly recent development in product design is the concept of recyclability. A product is designed so that it is easy to dismantle into easily recyclable or reusable components. A company may also choose to set up the recycling of its products so that the user returns the product after its useful lifetime. These aspects will incur further costs for the producing company and these need to be allowed for.

2.3.2 RETURNS FROM ENVIRONMENTAL ASPECTS IN PRODUCT DESIGN

Returns from a new product design can arise in several areas:

- sustained or increased sales;
- improved environmental performance and its effect on company risk;
- direct savings in production cost.

The identification of each of these returns will be discussed in subsequent sections. It is important to establish the purpose of a design change, because it indicates where the major portion of returns is expected and the analysis can be focused on this area.

In some cases a particular design change is required so that the product is allowed to be sold or so that sales can be sustained at the present level. It is difficult to put a value on these returns. Many companies do not include them in the analysis as the change in sales due to this specific aspect cannot be verified afterwards.

2.3.3 RISKS DUE TO PRODUCT DESIGN

Specific risks in the product design area arise from changes in consumer requirements and the regulatory environment. It is important to try to identify the expected consumer and regulatory reactions to a new design. The marketing department can offer advice on expected consumer/user requirements even though the product is not an end-product itself. A close co-operation with major customers can also provide valuable information on potential changes in sales volume due to design changes. The regulatory environment needs to be monitored by the department or function responsible for the company's environmental policies, so that an assessment of any potential effect on product design or packaging can be made.

2.4   PRODUCTION PROCESS

2.4.1   COSTS OF ENVIRONMENTAL ASPECTS IN PRODUCTION

Production aspects are summarized in Figure 2.5. A company may alter its production process to improve its environmental image, or it may need to make a decision about adopting a totally new production process for a specific type of product. In either case the operating costs of the process and the overall production costs will be different from the existing situation. The overall investment in the production process includes elements such as:

- raw material and energy use;
- waste treatment;
- production equipment;
- labour;
- maintenance;
- insurance on various potential liabilities;
- overheads — general overhead, quality, safety, health, reporting and so on;
- engineering and development;
- production site;
- material management.

All these elements contribute to the environmental performance of the production process.

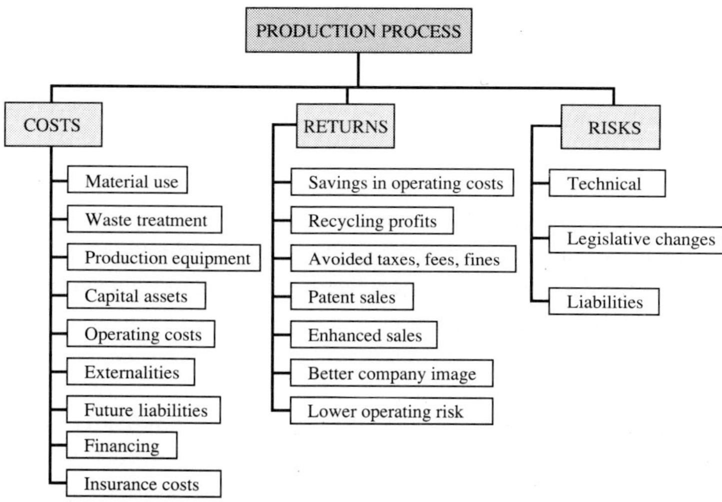

Figure 2.5   The production process module.

MATERIAL USE AND WASTE TREATMENT

Raw materials can be chosen either from renewable or non-renewable sources. Depending on the choice of raw material the production cost varies, as does the transportation of the material to the production site. Various raw materials produce different by-products that all have differing disposal requirements and selling potential. Different raw materials require differing pretreatment and different energy levels are used in the process. The choice of raw material is crucial not only to product quality but also for many other aspects of the production process.

A process also uses other materials that are not classified as raw materials but are nevertheless necessary, such as solvents, catalysts and cooling water. Some of these substances can be reused and recycled within the process. Others form waste, new compounds and by-products that have to be disposed of. Waste treatment cost should be included in the overall costing of the process. By-products can be either waste, or products that have a use and a market of their own. If these substances can be recycled, the operating costs of the process decrease. New compounds can be difficult to predict and therefore the cost of their disposal is not always easy to evaluate. Wherever incineration provides a feasible disposal alternative, its cost can be used — at least initially — in the investment calculation.

When contemplating various process alternatives the choice can be between a 'clean' technology and a technology that requires end-of-pipe treatment of the process waste. It is necessary to include the cost of waste treatment with the cost of those processes that are not inherently clean. Only then can a meaningful comparison of various process alternatives take place.

Determining the cost of various waste treatment options may prove difficult, particularly when the site is complex, demanding additional infrastructure and employee training. Some facilities are shared between products produced at the same site and it may be difficult to determine the operating cost for each product or product group. Experience has shown that contractors can determine the capital cost and operating cost of the isolated equipment relatively accurately. Less often do they have knowledge of the overall implications of a waste treatment option for the company.

ENERGY USE

When the overall use of energy is calculated through the lifetime of a product, energy use for transportation to the site and from the site is included. Also the energy used to process a raw material is included in the equation. From a company's point of view the energy used in getting the necessary raw materials, equipment or labour to the site is not a separate cost item.

24

The cost charged to a company, be it the total raw material cost or worker wages, usually also includes the transportation costs. In this aspect an investment calculation differs from typical LCA calculations. Only the so-called direct energy should be taken into account in the investment analysis. This is the energy needed in production. All other forms of energy can be ignored if a company is not charged separately for them.

PRODUCTION EQUIPMENT, CAPITAL ASSETS AND RELATED COSTS

Complications in this area mainly arise from non-engineering aspects, such as changes in the economic environment that affect currencies and the overall financing costs of a company. Also the estimation of the end-of-life value of production equipment may be difficult. Different industries have their own rules of thumb in this area. Valuation of the scrap value of waste treatment equipment is still unsettled.

Process changes require an input both from the R&D function and the engineering function. The breakdown of the R&D cost into various cost components is discussed in Section 2.2 (page 10). Engineering costs need to be defined similarly, particularly as the process development towards a better environmental performance may require substantial effort. Part or all of this cost may be external. Although the sum is easily identified, it is important to break down the overall sum to its components.

Environmental legislation is increasingly influencing the decision on where production should take place. Although many multinational companies are adhering to legislation in all countries of operation and do not wish to change or circumvent this policy, complex environmental legislation does accrue costs to a company. These costs arise in the form of the reporting and monitoring required by authorities. In most cases they are part of the overall manufacturing overhead. As the costs may be substantial, it is necessary to identify them separately and include them in the relevant investment calculations.

Many process developments take place either within existing production processes or existing plants. Therefore there are already many old assets in place and many facilities are shared by the whole plant. These may require alteration, expansion or curbing. It may be difficult to allocate operating costs of common facilities to various processes and also to detect the cost of changing the use of existing facilities. These costs may contribute significantly to the fixed cost base of a site and therefore should not be ignored.

OPERATING COSTS AND OVERHEAD

Direct labour cost is losing its importance as a cost component in most industries. Labour is mainly used for maintenance and overhead activities, such as quality control, process-related reporting internally and externally, and safety assurance. A new process or process change may require extensive retraining of staff, and even relocation for some. These activities are usually pooled together both in company cost accounting and investment calculations. This tends to disguise these costs items so that they are not attributed to the right projects and processes. Activity based costing is a way of breaking down the large overhead cost pools that many companies still have. In investment appraisals it is necessary that all overhead costs are correctly allocated to each process.

In many companies, there is a normal overhead rate to be used even in investment calculations. Unless it is clear what the rate covers, it is potentially dangerous to use it. Overhead definition can cover many vital categories of environmental costs, particularly those relating to waste treatment and monitoring. In external reporting of the company finances this may be the correct treatment. In internal investment calculations there is always a danger that some large cost elements may not be fully placed on the investment but diluted through the overhead allocation mechanism.

Transportation and storage of substances can result in an environmental hazard for which a company must take precautions. This may cover insurance, special transportation and storage vessels, special arrangements in the storage space and special attention to the safety of employees involved with the substances. All these precautions carry costs that need to be identified and included in the investment appraisal of a process.

EXTERNALITIES AND FUTURE LIABILITIES

Present implementation of LCA has led many companies to include externalities in their investment appraisals. Many of the externalities are the societal costs of losing — for example, clean air or water, clean nature and open space. Although these are serious considerations from society's point of view, they do not always represent an out-of-pocket expense for a company — unless a tax, such as carbon dioxide tax, is imposed upon companies, or they create a future liability for the company.

The cost of explaining why these substances are used and subsequently emitted is borne by a company. Many emissions are still allowed into the air, water and land. When treatment of these emissions is not required by law, it is up to individual companies to decide whether to undertake treatment or not. LCA provides valuable guidance about what emissions may take place. This inventory can then guide an individual company to make decisions on what it wishes to clean and what it should be cleaning.

Often not all the implications of the emissions from a production process are known. Some emissions may not be toxic or harmful in the short term or as separate substances, but when entering the environment they may start accumulating in the food chain or form new, toxic compounds that may cause an unintended hazard.

A company may become liable for these accidents and events years after the actual emission has taken place, so it is necessary to acknowledge the existence of these risks and cover the company against them. The insurance industry can offer professional service in this area and put a price on the probability of being exposed to such liability charges. Insurance is discussed further in Box 3 on page 28.

FINANCIAL AND INSURANCE COSTS

The costs of financing an investment enter into the investment appraisal calculation in many ways. Most commonly there is capital equipment to be acquired, stocks to be kept and financial risks to be guarded against. In many large companies, financing — or the so-called treasury operations — is done centrally and operating units are charged either a nominal financing charge or none at all. This tends to disguise the cost of financing. It is necessary to estimate these costs, though, with the help of the treasury or finance department.

Insurance costs tend to get forgotten if insurance and risk management is carried out centrally in a company. Insurance contracts may be based on historical information and some broad estimate of acceptable level of risk and compensation. Yet insurance premiums should be based, and increasingly are based, on actual and properly estimated risk levels. Each new process situation requires a proper accident risk evaluation. This should be reflected in the overall insurance cost of the company.

HOW TO USE LIFE CYCLE ANALYSIS AND OTHER ENVIRONMENTAL MANAGEMENT TOOLS IN INVESTMENT ANALYSIS

The process of LCA covers the whole life span of a product from raw material extraction to the discarding of the end-product (see Figure 2.6 on page 29). The common steps in such an analysis comprise an inventory take of the use of environment's resources, evaluation of the environmental impact and planning of improvements. The process can be costly and time-consuming and it often presents many technical and philosophical difficulties.

Although not all detailed data is available for a full LCA, the technique provides a usable framework for an investment evaluator. The inventory analysis helps to highlight the most important emissions and largest burdens on the environment, whether they are intense energy use, highly toxic compounds in effluent from production, large emission quantities or costly waste treatment.

BOX 3:
Insuring Environmental Liabilities

Insuring environmental liabilities can be a hazardous business for the insurance companies. This is clearly demonstrated in a Financial Times *article of 13 April 1994 which reviews a study on the developments of Superfund legislation and its interpretation. The study was conducted by AM Best, a US insurance analysis and rating agency in New Jersey.*

AM Best calls the pollution and asbestos claims a 'black hole' for the insurance industry. The agency's report claims that the insurers need to reserve an additional $260 billion to meet their exposure to environmental and asbestos claims over the next 15 to 20 years. This is due to Superfund legislation in the US.

Superfund legislation is based on the concept of 'the polluter pays' for cleaning up the site or water body polluted. The interpretation of the law has created a problem that the manufacturing and insurance industry have to face up to. The actions of previous owners are also targeted under Superfund legislation independent of the actions of the present owner of the site. This has led to a chase after previous owners who should participate in the costs of the remediation programme.

The application of Superfund legislation has led to an increasingly unknown contingent liability for the insurance industry. The manufacturing and process industries will be the first port of call for funding the clean-up, but they may be able to claim the cost in their insurance. All this is happening after the premiums have been set, the production has taken place and the site may have been sold to a new owner.

British and European Union legislators are adopting a similar approach to responsibility for the cost of remediation. Although no Superfund exists, companies can be held retroactively liable for pollution they have caused. This is despite the fact that at the time of the emissions no regulations were in place to forbid them and the relevant permits had been obtained and followed.

The repercussions of these approaches can be lethal to many companies. First, insurers do not insure against gradual pollution. Second, companies have no way of knowing their future liabilities but have to be prepared to pay for anything. Third, this must have an adverse effect on the investment willingness of the industry. Has your company realized that it may have to take into account an average $31 million for clean-up costs that may be imposed upon its site — as has been the case in the US?

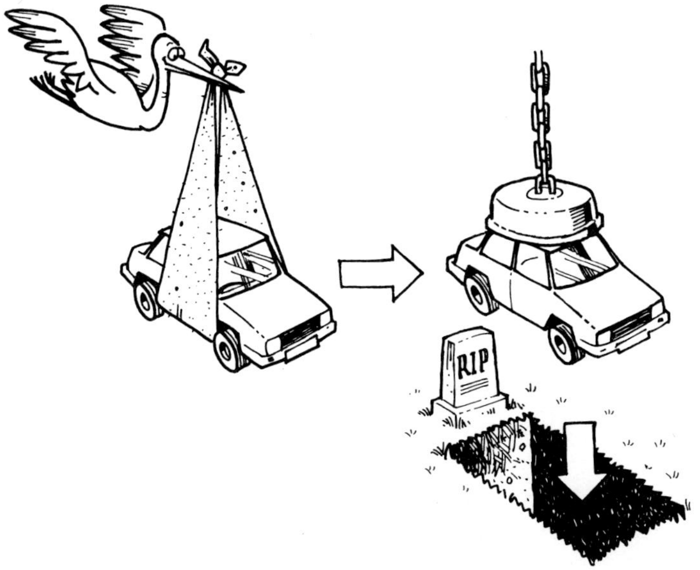

Figure 2.6   Life cycle analysis from cradle to grave.

The results of the LCA must be looked at initially from the company's point of view. The company should identify which issues in the LCA results entail most costs. The results may influence the consumer perception of the company and its products. They will highlight where problems arise in terms of allowed environmental loading and overall licence to operate.

An LCA on dishwashers is a good example. The LCA shows that the major burden of a dishwasher on the environment is not from its production but from its use. Therefore the emphasis of effort in improving the environmental impact of a dishwasher is placed not in production but in R&D and product design.

Other environmental management tools are public health assessments, Hazop/Hazan processes, TQM, BS7750 and other similar quality system standards, environmental auditing and environmental management systems. All these provide input into the investment appraisal. They highlight areas for potential future liabilities, additional investments required to achieve a specific environmental goal and costs and benefits to be gained from these investments.

INTEGRATED POLLUTION CONTROL APPLICATION PROCEDURE AND COST OF PUBLIC CONSULTATION

In England and Wales, Her Majesty's Inspectorate of Pollution (HMIP) looks to improve every production site from the point of view of integrated pollution control, so that the processes follow the principles of BATNEEC (Best Available Technology Not Entailing Excessive Cost) and BPEO (Best Practicable Environmental Option). Similar environmental permit procedures are in place in other countries. The process of preparing the applications for a permit or licence to operate may bring substantial costs to the company due to the commitment involved.

Often the permit procedure required officially is not adequate and a public consultation needs to be undertaken. In some countries there are legal requirements in place for the format of such consultation; in others the consultation can be undertaken in a free format. In both cases the process incurs the company costs, covering not only the actual documentation but also arranging the forum for the consultation process to take place.

2.4.2 RETURNS FROM ENVIRONMENTAL ASPECTS OF PRODUCTION

Process changes resulting in improved environmental performance can give returns from:

- material and energy savings;
- recycling profits;
- reduced or avoided waste treatment costs;
- cheaper raw materials and supplies required;
- implementation of integrated technologies;
- reduced or avoided licence payments and compensations, taxes, fees, fines;
- patent sales.

These were identified as the most common forms of returns in a German study conducted by the German Federal Environment Agency (*Umweltbundesamt*)[6]. It is evident that accounting in most companies cannot show with adequate clarity what the returns on process investments are. An American study[7] adds to this list the following returns:

- increased revenue from enhanced product quality;
- increased revenue from enhanced company and product image;
- reduced health maintenance costs from improved employee health;
- increased productivity from improved employee relations.

The first group of potential returns may be easier to estimate as they depend on the internal actions of a company. The second group of returns (identified in the American study) is based on the perceptions of consumers and employees. These perceptions, and particularly their impact on company returns, are difficult to estimate. The issues of company image and product quality are discussed in Sections 2.5 and 2.6 respectively, pages 32 and 35.

In the case of totally new processes and production methods, the returns mainly come from two sources: sales of the new product, and potentially improved environmental image of a company. The improved environmental image may influence the overall financing and insurance costs of a company. These returns affect not only the investment under study but the company overall.

The impact may also be delayed due to the review processes required from the external stakeholders. Therefore the impact may be difficult to incorporate in the investment appraisal of a project.

Process industry production is increasingly becoming the target for various pollution permit arrangements and related compensation schemes. Adoption of a changed process technology, a new end-of-pipe treatment technology or a totally new production process usually means changes to the permit arrangements and economic compensation. In some cases these may be significant and should therefore be included in the investment appraisal scheme.

2.4.3 RISKS IN ENVIRONMENTAL ASPECTS OF PRODUCTION

To an ever-increasing extent the risks associated with changes in production processes are related to changing environmental legislation and the liabilities that may result. The environmental liabilities may be so significant that they make a company insolvent, particularly when in some cases insurance companies are not required to compensate for environmental damage that the company has known about. It is therefore essential that the environmental consequences of production processes be studied carefully when an investment appraisal is being made.

Many of the environmental risks in production are due to potential hazards involved in the production process. Hazan/Hazop studies are designed to identify the areas where more stringent safety measures are required. Contractors and equipment suppliers conduct studies on the reliability of various unit processes and of piping, vessels and other equipment. These estimates can be used by the investment evaluator in assigning probabilities to environmental hazards.

Emissions into the environment can form new substances that were not expected or whose effect on the environment was not fully understood at the investment stage. These risks cannot be taken into account in the initial investment study and the company can only try to adhere to known legislation and scientific data.

Environmental legislation is becoming stricter in most countries. The pattern for this development is difficult to foresee. Many companies are reluctant to make an investment that allows them to outperform the legislative requirements, as this may set a precedent for the regulator. Potentially, the company can make an internal estimate of how much more investment in waste disposal or treatment is needed to achieve a particular decrease in emissions of a substance. Such a cost buffer could be included in the investment calculation.

2.5　MARKETING AND THE MARKETABILITY OF PRODUCTS

Designing and producing products with a positive environmental performance and image may be based on various objectives that a company wishes to reach:

- the overall production cost is decreased and hence a better sales performance is expected due to a cost and/or price advantage;
- a product or process innovation is exploited commercially;
- a particular niche of the market is tapped in the hope of extraordinary gains;
- an environmentally positive image is believed to improve sales and company image;
- risk of environmental liability claims is decreased.

This book concentrates on process industry products with a 'green' image. Although it is true that many product and process innovations in the process industries may have markets of their own as technologies, these are not discussed here. Process industry products are both intermediates and end-products sold directly to the end-user. Both types of products are discussed here.

Although many process industry products are intermediates, consumer expectations play an important role in the sales success of products. In the case of an intermediate product, consumer and regulator expectations are crucial: a product, such as CFCs, cannot be produced because of regulations. Many chemical industry products have to be recyclable in order to be sold at all. End-product regulations and consumer expectations can either stop sales of a product or seriously restrict them. Detergents containing phosphates are an example of this.

Marketing issues are summarized in Figure 2.7.

2.5.1　COSTS IN SELLING AND PRODUCING ENVIRONMENTALLY-FRIENDLY PRODUCTS

Investments in R&D, changing product design and/or production processes have an impact on the marketing and the marketability of a product. This is demonstrated by the numerous regulations that are in place for products and the characteristics of those regulations. Other factors that influence the overall investment in marketing and the marketability of a product are product promotion, market studies, market creation and lobbying.

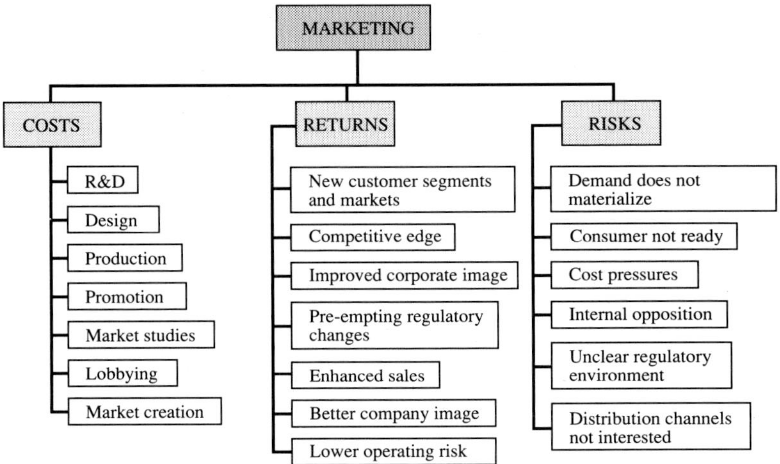

Figure 2.7   Marketing and marketability of products.

Market studies are of particular importance because they indicate whether a green image brings any extra sales — that is, whether consumers have any environmentally-geared expectations about the product. Market creation refers to the efforts of a company to promote the green image of a product actively, thereby creating a demand for it. A company or an industry may wish to ensure that its products in a category can be sold, or no greater restrictions are placed upon them.

If a product is changed to become environmentally 'friendly', it may be necessary to find new distribution channels. The cost may differ drastically from that of the existing channels used by a company. Marketing must also be redesigned and the company incurs more costs here too. This can be particularly true if the company seeks an eco-label for the product. These labels are not restricted to end-products, and products like pulp can also receive them. Eco-labelling schemes and the related costs are discussed in Section 3.4, page 56.

2.5.2   RETURNS FROM ENVIRONMENTAL ASPECTS OF A PRODUCT

Returns from an environmentally-friendly product (that potentially has also been produced by an environmentally-friendly process) can arise in several different ways[8]:

- gaining new customer segments with good purchasing power;
- reducing the gap between the product and competitors' products;
- realizing a long-term competitive edge;
- improving corporate image;

- gaining cost savings by pre-empting regulatory changes;
- finding new markets;
- improving co-operation with distributor organizations.

The general expectation in companies on both sides of the Atlantic is that a green image brings extra sales, although some 30% of German companies interviewed in 1991 still doubted the sophistication of consumers[9]. The problem is quantifying these 'extra' returns, particularly in the case of an intermediate product, such as a solvent, pulp or CFCs. In the case where a product can or cannot be sold due to regulations, quantification is clear-cut: markets where restrictions apply can be ignored.

To evaluate the consumer response to a green image a market study must be conducted. For intermediate products, client companies can provide valuable information. For end-products, reactions to existing eco-labels in similar products can be used as a guideline. An important aspect in testing consumer response is the price of the product being introduced. A positive environmental impact may be desirable but the consumer may not want to pay a price premium for it. Instead a larger market share may be gained. Success of eco-labelling schemes is discussed in Section 3.4, page 56.

Conjoint analysis (CA) has been around as a market research tool for a few years. Here, consumers make choices between different bundles of product characteristics as opposed to single, individual characteristics. By forcing consumers to give up some characteristic or a certain amount of it in order to gain another characteristic, the analysis can highlight the desirability of the environmental aspects of a product compared to other features.

2.5.3   RISKS IN SELLING ENVIRONMENTALLY-FRIENDLY PRODUCTS

Marketing a product is based on some particular product characteristics and the environmental performance of a product is a further example. The long chain of decisions on what aspects of a product to promote begins with R&D, product design and process design. All the choices in these areas have been made, or should have been made, on the basis of the desired marketing characteristics of the final product. The decisions in various areas are based on the expectation of development in the regulatory framework and the expectations of consumers. If these change, the future of the product can be jeopardized.

Building the marketing of a product on environmental aspects presents potential dangers[10]:

- demand for environmentally-friendly products does not materialize as expected, or fades away;
- the consumer is not ready for environmentally-friendly products because of technical difficulties;

- there is pressure on cost due to environmental investments;
- environmental performance is not adequate on its own to improve the competitiveness of a product;
- there may be opposition within the company to promoting environmentally-friendly products or production processes;
- the regulatory environment is unclear and provides uncertain guidance for product development and marketing;
- the distribution channels are not interested in the product.

All these dangers underline the importance of co-operation within a company. The marketing function plays a significant role in informing other functions of the expectations of consumers and the potential changes in them. Although changes in the regulatory environment are closely linked to the expectations of consumers, monitoring of legislation should not be left to marketing staff alone. Interpretation and monitoring of cases which set precedents requires legal expertise. This is particularly true when new legislation is being interpreted for the first time. For instance, in the USA experience has shown that interpretations of the law have taken unexpected turns.

These risks can be quantified by normal market research techniques. Conjoint analysis, referred to earlier, provides a means for making monetary estimates of the impact of a characteristic on sales and price. The risks in this area can be reduced by doing the homework properly in advance, studying the market and the distributors thoroughly.

Test Case A (page 63) shows the importance of market research. The whole project was based on the assumption that there would be an enlarging and profitable market for chlorine-free pulp. The way the company hedged itself against an adverse development in the market place was to phase the development and production of the new product over some years. These years then hopefully showed the direction the market is going to take. This approach avoids the problem of having to incorporate market risk in the investment calculation, and the hedging cost is implicitly taken into account in the size of the investment.

2.6   CORPORATE IMAGE

Most companies believe that a good corporate image is essential for their success. 82% of companies interviewed in a survey believed that a good environmental performance image has a positive impact on their corporate image[9]. It is much more difficult to define which factors create corporate image, how to influence these factors and how corporate image is translated into profits or losses. Box 4 on page 36 gives an example of this issue.

> ### *BOX 4:*
> ### *What About Company Image?*
>
> *Do you remember what happened to the Rhine in 1986? Even if you had no interest at all in environmental issues, you would probably still remember the story of Sandoz. The people of the town of Schweizerhalle remember so well that they were able to delay the planning permission of a special waste burning facility in Basle until 1991. The project was initiated in the early 1980s.*
>
> *The case of Sandoz is a typical story of a single accident giving the site where it occurred and the whole industry many problems. In the early morning of 1 November 1986 a fire broke out in a warehouse and sent acrid smoke over the city of Basle. Some 30 tonnes of dangerous chemicals, including 200 kg of mercury, were released into the Rhine. No human lives were lost, but the marine life in the Rhine suffered badly. Some species, such as eels, were killed off completely.*
>
> *An investigation into the accident revealed no negligence on the part of Sandoz. Simply, no one had foreseen what might happen in the case of a fire at the warehouse. In particular, no one had thought what would happen to the toxic liquids forming in case of a fire. Yet the relationship between the company and the local population deteriorated to such an extent that the company was unable to get planning permission for any project for years.*
>
> *This accident has had an impact on the chemical industry as a whole. The Responsible Care programme has been launched, Hazan/Hazop studies are more a practice than an exception today, and many companies — such as Norsk Hydro, BP, ICI, Rhône-Poulenc and Ciba-Geigy to name a few — now publish their environmental reports.*

Corporate image is a result of actions taken by a company, the communication policy pursued and the strategy followed. In the case of environmental performance image, several factors contribute:

- the strategy adopted by a company and the way it is communicated to the employees and other stakeholders;
- the overall communication policy adopted by a company in cases of hazards and accidents;
- the actual performance and the perception of it by stakeholders;
- the marketing strategy adopted and its application in a product portfolio.

The problem from a company's point of view is that the corporate environmental image depends almost entirely on external perception, because reliable and easily interpreted measures of image do not yet exist.

Work has been done on developing various types of eco-balances, emission sheets, energy counts and the like, but these have raised more questions than they have answered. Some companies, such as Norsk Hydro, have already published an eco-balance and others, such as ICI, make a separate annual report on their environmental programme. Such actions are very much at the discretion of the company, however, as no exact, legally binding rules yet exist about how to inform stakeholders on the environmental performance of a company.

Financing cost and insurance premiums are two categories of expenses that may depend largely and directly on the environmental image of a company. Due to the interpretation of legislation in different countries, lenders may get more or less involved in environmental liability issues. In some countries, such as the USA, lenders may become partially liable for environmental hazards. In most other countries, the risk for a lender is the financial viability of a company that has a great potential for future environmental hazards and liabilities. The banking community is increasingly taking these issues into account when extending credit.

Insurance companies and national insurance associations are in the process of deciding how to insure companies with potential future environmental problems. In each country very different approaches are being adopted, ranging from total refusal to insure against certain risks to situations where emission permits cover companies against most risks. Depending on the legal framework, insurance costs play a more or less significant role in investment costs. It is necessary, though, to evaluate the impact of new processes or process changes and products on the future environmental liabilities of a company. These may then be covered by different types of insurances and at differing costs. Issues of corporate image are the subject of Figure 2.8 (see page 38).

2.6.1 COSTS OF CREATING AN ENVIRONMENTAL IMAGE OF A COMPANY

Investments made in improving the environmental image of a company are at the discretion of each company; it has to decide on the policy for communicating aspects of its environmental performance. Hence environmental communication becomes a part of the overall promotion a company undertakes. Various promotional tools are available to a company including:

- publication of research results concerning a product;
- publication of LCAs;

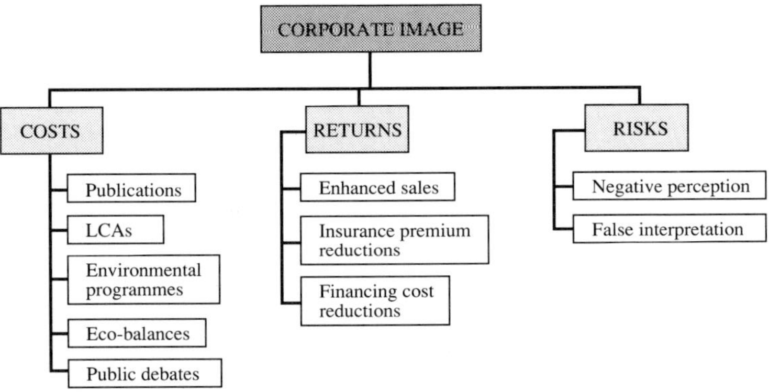

Figure 2.8   Corporate image.

- publication of an environmental policy and programme, and the achievements within the programme;
- publication of eco-balances;
- promotion of environmental aspects of the product portfolio;
- taking part in public environmental debate.

All these investments require a strong strategic background; a company has to have an environmental policy to be credible in its efforts. All these investments accumulate a cost to the company, and have an intrinsic risk if the company cannot live up to the expectations created through publicity around its environmental goals. These investments may lead to other investments in the company. Monitoring equipment may have to be installed and production processes altered.

Test Case B (see page 88) reveals the vast investments that are made in production plants for ensuring public approval of operations. 30% of the operating costs of the plant were accrued by a process that did not improve the actual environmental performance of the plant or of its product. Yet the company, ICI, initially deemed that it had to make this investment in the interest of public image. Later, these process changes were required by the regulator.

The two groups of stakeholders that may require special attention are the insurers and the financing sources of a company. These sources may take sizable risks in a company and therefore want to be well informed about future potential risks. Close co-operation with these stakeholders may be beneficial to a company in the form of cost reductions.

The cost of promoting an environmental image may be difficult to calculate for an individual product. It is more likely that the cost will accrue to a group of products, such as plastics or fertilisers, or to a unit process, such as bleaching of pulp. Often the expenditure may have to take place long before the benefits materialize. This is particularly true with insurance premiums that are reviewed on the basis of historical accident records. The historical nature of these costs may also mean that they are sunk costs in an investment appraisal situation. Yet these effects may be significant enough to require an estimate at the investment appraisal phase.

2.6.2 RETURNS FROM AN ENVIRONMENTAL IMAGE

Companies frequently undertake campaigns to inform the public and stakeholders of their environmental goals and achievements. Eco-labelling schemes are very popular. Environmental aspects of products are highlighted in their promotion. Thus companies obviously think that there is a return on the investment in this promotion. In the case of eco-labelling some clear results of sales success have been collected. For general promotion of company image, however, results are more ambiguous, if they exist at all. Particularly for the process industries, which are not always visible to the end-consumer, the need for heavy environmental promotion may be questioned.

A company should establish its existing image before undertaking a promotion campaign, and then determine what it wants to achieve. Marketing and public relations professionals can then make an estimate of the costs required to achieve the goal. A company should also evaluate and analyse the factors that form the public image of its particular environmental performance. Only then can the company fully evaluate the returns on the investments it makes in promoting a good environmental performance image, be they in R&D, product design or process changes.

The returns of a good environmental image may be difficult to attribute to a specific product. There are two ways the returns can be more easily defined. First, the insurance premiums paid at a specific site may depend on the overall image of the company. If a company is perceived to follow good environmental management practices as indicated, for instance, in an environmental audit, the insurer may feel less need to add an additional risk premium. Second, the banking community needs to assess the environmental risks involved in a company. A good image may reduce the need of an extra risk premium and hence bring down the financing cost.

2.6.3 RISKS OF AN ENVIRONMENTAL IMAGE

A company should also be aware of the risks of a poor environmental image. An environmental hazard can affect the company's image in a negative way. A minor leak is fast forgotten but market studies have established that an accident like the Sandoz leak to the Rhine harmed the image of the whole chemical industry in Europe for a long time. A poor environmental image may mislead consumers to adopt practices which are actually more environmentally harmful. A typical example of this is the debate on PVC versus other packing materials. PVC bottles for soft drinks are often regarded as less environmentally friendly than glass bottles. Yet glass bottles require recycling, washing and more transport per litre of liquid and, therefore, are actually more harmful to the environment than PVC bottles.

The tools used to promote a good environmental image may turn against a company. Eco-balances and the like do not have a universally accepted methodology and therefore pressure groups can actually use them against a company. A company can be accused of bias and false presentation of facts. Evaluation of environmental performance is always done against the benchmark of what the present legislation requires or what the expectations of various stakeholder groups are. The danger that this presents to a company is the variability of these expectations and the rapid changes in them. A company must understand what is expected from it and adjust its image campaign accordingly.

In determining the monetary value of these risks, a two-tier approach is required. First, the company needs to measure its existing company image and gain data on how various measures affect its image. Second, the company needs to measure the financial implications of an image. Market research techniques, such as conjoint analysis, can help in this task.

It is clear that studies to determine that impact of company image on sales and licence to operate require significant effort. This may lead companies to reject such studies as unnecessary. Yet without any further knowledge, companies are prepared to make significant investments in improving their image. Which other investments are made without any understanding of the income prospects?

2.7 OTHER ASPECTS OF THE METHODOLOGY

The methodology described in this book encompasses several fields of expertise in a company. It also builds on some technical requirements without which its use is not possible (see Figure 2.9). These include:

- risk and risk modelling;
- discounting methodology;
- consideration of a deterministic versus a probabilistic model;
- cost accounting issues;
- ways to use the methodology in practice.

They are discussed in detail in the following sections.

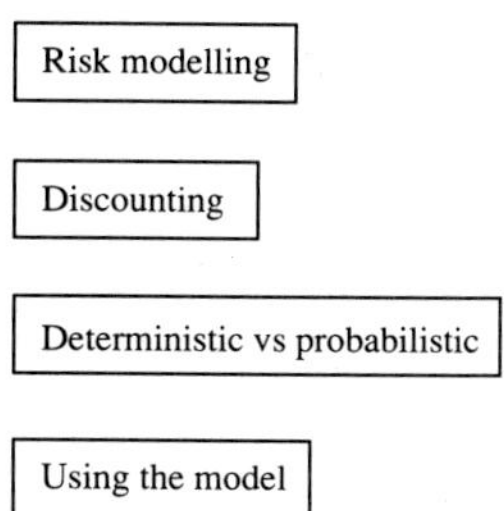

Figure 2.9   Other aspects of the model.

2.7.1   RISK MODELLING

EXPECTATIONS OF STAKEHOLDERS

Investment calculations are made in an uncertain environment. There can be statistical information available that can help in quantifying the risk. There may be some evidence that can be used for estimates, but many of the uncertainties that are relevant in this case can only be evaluated and estimated by an informed person.

The uncertainties arise from:

- changes in external and internal stakeholder expectations;
- technical aspects of the investment under consideration;
- the conditions of the markets where a company operates.

External stakeholders comprise many different groups: pressure groups, regulators and governmental bodies, transnational organizations, consumers, the medical profession, banking and the insurance industry. All these organizations have opinions, expectations and information that influence the conditions in which a company operates.

The internal stakeholders are employees and owners of a company. These groups also have varying expectations on the environmental performance of a company. Although these expectations may follow the overall pattern in society, the involvement of internal stakeholders in the company may influence their expectations as well.

The problem is that expectations change over time and often in an unpredictable fashion. The actions of pressure groups influence political decision-making; the medical profession makes a new discovery that influences legislation and trade unions; both the banking and insurance industries are becoming increasingly aware of the potential risks of environmental hazards and accidents. This is reflected in the risk premiums added to interest rates and insurance fees.

TECHNICAL AND MARKET-RELATED RISKS

Risks of a technical nature are connected with the production processes, R&D, relative costs of inputs into production and the cost of waste treatment. The economic environment dictates the return on an investment by altering the cost of both operations and financing through changes in exchange rates and interest rates. Trade agreements between countries influence the terms and conditions of trade. Artificial trade barriers may be built between countries. All these aspects have an impact on the overall competitive situation of a company.

The markets in which a company operates may undergo changes that increase the risk to a company. Changes in consumer tastes have already been mentioned. It is also possible that new, more price-competitive products enter the market. Investments in environmentally-friendlier technologies and waste treatment may increase the cost of production and thus undermine the price competitiveness. Company image can suffer due to an accident either in the company or in the industry the company represents.

The cost of the required inputs to production may vary over time and in relation to each other. If one of the inputs is a raw material with a world market price, such as oil, the price fluctuations can be large. Although in many cases there is a possibility of using commodity market-based options or futures to provide some protection for a project from adverse price fluctuations, it is necessary to build various scenarios where the cost levels and internal relationships vary to get an idea of the sensitivity of the project to cost variations. If commodity options are used, the cost of these must be taken into account in the investment calculation.

An analysis of market conditions needs to be conducted to account for the risk of adverse changes. The likelihood of new competitors entering a market can be estimated on the basis of barriers to entry (such as large capital cost or trade barriers in legislation). Pricing of competing products should be studied carefully to estimate whether a product can be sold at the required price. If a product cannot be sold in a particular market due to restrictions, environmental hazards harming company image or for some other reason, it is necessary to estimate the cost of having to transfer sales from one market to another. The probability of such an event will be established by a subjective estimate from an informed person. If a company feels uncertain about the probability, a low percentage can be assigned to the event. It is important that a cost element is included.

Table 2.1 shows a model of risk assessment applied to analysis of effects.

TABLE 2.1

Risk assessment applied to analysis of effects (Source: Hunt, D.T.E., 1994, Management of environmental matters, *Manenpro '94* (IChemE).)

THE BASIC APPROACH

(1) For each effect, assign a rank order (say 1–5) in respect of:

- frequency of occurrence (F);
- likelihood of control loss (L);
- severity of consequences (S).

(2) Multiply these to obtain an overall criticality factor (C):

$$C = F \times L \times S$$

(3) Rank effects by their C values and judge significance accordingly.

ASSIGNING FACTOR VALUES

- Frequency of occurrence (F)

1 = Very rare — for example, complete failure of a robust 'fail-safe' system — to ...

5 = Continuous — for example, a treated effluent discharge

- Likelihood of control loss (L)

1 = Very unlikely — for example, loss of control of a robust control element — to ...

5 = Highly likely — for example, loss of control of biological treatment system

- Severity of consequences (S)

1 = Very limited, localized impact — for example, local odour problem — to ...

5 = Extensive and severe damage — for example, toxic spillage to major river

FUTURE ENVIRONMENTAL LIABILITIES

The expectations of insurers and bankers also have repercussions on the financial outcome of the model. Increasingly both industries are demanding environmental audits to be conducted before any services are offered. In the US many insurers are refusing to extend insurances without an exclusion clause for environmental liabilities, or else the insurances are for one year only. This leaves companies in a very exposed position and underlines the importance of including the probabilities of environmental hazards and liabilities in investment calculations.

It is suggested that LCA is used as a source of information for determining emission quantities. Although the financial value of a liability claim due to emissions is usually unknown due to lack of precedents, the clean-up cost and the extent of a hazard can be evaluated with some precision. The technical probability of a hazard can be evaluated on the basis of studies of, for instance, container safety and process safety. This type of work has been conducted in the USA by the Environmental Protection Agency which has devised a so-called Hazardous Waste Tank Failure Model. Although the estimate is crude, it is important to include it in the project appraisal: the financial repercussions can be substantial.

The extra costs of financing due to risk premiums are difficult to evaluate. An approximation of the risk premium can be, for instance, the cost of an environmental audit and information requirements set by the financing institutions. General Electric has developed its own approach to evaluating the future liability costs of waste management options[11]. It builds on steps for structuring a risk management model. First, the evaluator ranks the risk of various waste management processes. Then waste management facilities are ranked utilizing risk factors. Finally, the volume of the waste stream and the concentration of specific hazardous constituents are estimated. Reduced volume is reviewed as reducing future liability risk. The 'GE handbook' provides help in estimating future liability costs[11].

INCORPORATION OF RISK IN THE ANALYSIS

All these risks need to be taken into account when estimating the costs and returns of an investment, whether it be in a process technology, R&D or product design. Two elements are needed to incorporate these risks into the investment appraisal. These are a formula for calculating the risk and sufficient data to put into the formula. Information on the trends of external expectations can be obtained from various sources. For example, trade organizations are conducting surveys among various groups; pressure groups produce publications that often indicate the direction for legislation; client contacts and market surveys offer information on the expectations of consumers; medical journals concerned with occupational health provide information on recent research. Both the banking and insurance industries are increasingly setting up their own environmental departments and formulating policies for handling environmental risks in companies. Internal expectations are often expressed by trade unions, company policy documents and in shareholder surveys.

These changes and expectations can be incorporated in different ways. The expectations may influence company policies in such a way that certain emission types are not acceptable or certain pollution types are being restricted internally. These internal targets need to be met or the investment is not considered. Usually, expectations influence the product portfolio of a company and the expected returns on the various products. When estimating the sales volume a probability should be assigned to the expected sales figures. The choice of probability distribution and respective probabilities for each possible outcome of each parameter is subjective and the decision of an informed person.

Technical methods of modelling risk vary greatly depending on which particular module of the model is considered. Engineering sciences have developed many techniques that are gaining wider use, such as event and fault tree analysis. These help the analyst to break complex systems down to their components and analyse each component separately. Monte Carlo simulations are used increasingly in various business situations. This has been made possible with the introduction of easy-to-use and inexpensive PC applications.

Although the techniques exist, they are not foolproof. The quality of results depends very much on the ability of the analysis team. Problems that may arise include:

- how to treat common cause problems and hazards that arise from a single event;
- how to allow for increase of uncertainties in primary outputs;
- how to ensure a sufficient degree of comprehensiveness of the systems definition;
- how to specify the boundaries of the decision trees developed so that the problem is numerically tractable;
- what form and parameters of probability distributions to incorporate in the data inputs.

The methodology that is presented in this book aims to define the system boundaries and give ideas on parameters and factors to be taken into account. It further aims to give ideas on how to incorporate risk in various situations and which alternative measures can be used to replace the ones that are difficult to quantify. Some guidance is also offered on where to find the relevant data. Each analysis situation is very different due to the nature of the technology and the product under consideration. The analysis may reveal a high level of uncertainty and a high risk of an occurrence that may jeopardize the whole company. However, the investment may be acceptable from a strategic point of view.

2.7.2 DISCOUNTING ISSUES

The time value of money needs to be considered in an evaluation of project profitability. This is particularly important when a project has a long life span or is conducted in inflationary conditions. Discounting — that is, transferring all relevant project cash flows into the same time period — is commonly used in traditional investment analysis. The aspects of discounting that need special attention are:

- the inclusion of risk;
- identification and definition of relevant cash flows;
- determination of the discount rate;
- determination of the time horizon of the project.

Many of the variables to be included in the project appraisal carry an inherent risk. Until recently the most common way of including risk in the equation was either by building in contingencies or by increasing the discount rate and the required hurdle rate for project acceptance. In many cases risk was double-counted, and no sophisticated models for assessing the probability function of the various variables were employed. Today the introduction of PC-based risk management and modelling tools makes production of probability distributions for a number of variables easy. This allows inclusion of risk to be straightforward, so that contingencies and abnormally high discount rates and hurdle rates can be abandoned.

One of the basic assumptions of traditional investment analysis, such as net present value (NPV) calculations, is the inclusion of cash flows only. Two major flaws can be detected in numerous published investment calculations. Depreciation is included even though it is not a cash flow (although it is correct to include depreciation when making an after-tax calculation). Working capital is included at its asset value and not at its financing value. Working capital requires financing, which in turn causes interest payments that are true cash outflows. These are the ones to be included in the appraisal. Although in some cases balance sheet values are used in NPV calculations, it is important to realize that mixing different types of parameter definitions will produce a faulty result. The extension of cost and revenue categories in the appraisal has been discussed in each module of the methodology.

Selection of the discount rate is problematic, particularly when a company is operating in a high inflation environment. The basic rule is that the discount rate should mirror the financing cost of the company and hence the overall financing risk that lenders and the stock market perceive. It is therefore necessary to calculate an approximate weighted average cost of capital to a company. This can then be used as the discount rate. The capital cost also mirrors the risk perception of the financial markets and the rate at which cash can be reinvested in the money/capital markets during the lifetime of the project. The discount rate should not include any premium for project risk, because the various elements of a project contribute a different level of risk.

Most companies apply decision rules for an acceptable return on an investment. This rule can be a hurdle rate for either internal rate of return (IRR) or the net present value (NPV) of a project. Typically, a project should also have a particular payback time within which the income from the project equals the investment made into the project. In recessionary conditions, many companies wish to shorten the payback time required from projects in order to accept the investment. Projects which improve the environmental performance of a company may suffer unduly if the time horizon for project appraisal is very short.

Most environmental benefits and costs amass over time. How long a time horizon is allowed in investment appraisals is a matter of company policy, considering that environmental costs and benefits may occur over a long time span.

Each company has its own guidance on acceptable time horizons of investment appraisals. The first guideline for selecting the time horizon is the expected physical life of an investment. If the investment involves capital assets, there may be a need to assign a scrap value to them. Another useful guideline is to evaluate the real impact of cash flows far into the future under different discount factor ranges. For example, if the discount rate is 10%, cash flow after 7–8 years becomes very insignificant. A potential approach is to convert these flows into perpetuities.

2.7.3   DETERMINISTIC VERSUS PROBABILISTIC MODEL

The most commonly-used versions of investment calculations are deterministic: risk and uncertainties are not explicitly considered. A verbal comment is often added to the appraisal statement. Even more often risk is buried in the discount rate — incorrectly, as argued above. A deterministic model is acceptable if the cash flows under consideration are fairly certain, inflation well estimated and the overall operational environment is fairly stable.

For investments with an environmental impact, uncertainty is inherent in the analysis. The uncertainty arises because emissions from a process or a product enter into the biosphere and the reactions are not well known either in extent or timing. Additional complications arise from the fact that the outcome of an investment project is greatly influenced by human behaviour, such as the development of expectations.

Many of the parameters included in the analysis of a project outcome are based on an unknown probability distribution. It is therefore important to attempt to make an educated guess at what their probable value may be, since they have a significant impact on the profitability of an investment. The best example of this is liability costs. Experience gained in the US demonstrates clearly the gravity of the liability concerns.

These aspects point to the use of a probabilistic appraisal model and many of the processes for which the model can be used as a basis for project appraisal are technically very different.

2.7.4   COST ACCOUNTING ISSUES

A problem commonly encountered is that company overheads may encompass a great variety of costs that could be attributed to projects and products. When identifying such costs, it is advisable to review them with the accounting department which can help identify costs needed for the project appraisal.

Project appraisals are typically done for a separate project and not on the basis of comparing alternatives. The comparison is on the basis of hurdle rates, payback times or IRRs and in many cases this is an acceptable approach. For investments in clean process technologies, however, this approach may be inadequate and could lead to unnecessary discarding of profitable projects.

The situation can be illustrated by an example. A company may be contemplating whether to invest in a cleaner process-integrated technology or keep the old process and add an end-of-pipe treatment. It would be wrong to compare the investment cost of the end-of-pipe facility with the total cost of the new technology, as the full impact on product cost is not revealed. Two possible approaches are comparing the two processes on the basis of resulting product cost, or regarding the avoidance of a waste treatment facility cost as an income to the new technology investment. Thereafter operating or resulting product costs can be compared realistically.

A comparison can be made of a minor alteration to a process versus building an end-of-pipe facility on an incremental cost basis. Instead of analysing the full cost of resulting products under the two scenarios, the incremental cost of improving the process and the incremental cost of adding an end-of-pipe treatment can be compared. The difficulty with this approach is that many costs may not be recognized.

When comparing two different investment alternatives, the use of opportunity cost and return are recommended. If a company wishes to stay with the existing process technology, it should analyse the opportunity costs and returns arising from this 'investment'. A cleaner technology can reduce a particular emission type and thus reduce waste treatment costs and liability risks, leading to an 'income' to the company. If the company stays with the old technology, this income will not materialize. This is an opportunity loss to the company and should be considered a cost of operating the old technology.

2.7.5   USE OF THE METHODOLOGY

The methodology presented in this book can be used in many ways. It helps to analyse the modules and areas in company operations that are to be considered when making an investment appraisal. It attempts to demonstrate the needs of various parts of the company organization to co-operate in order to bring together all the necessary information for project evaluation.

The method is complemented with a series of 'check-lists' that can be used when analysing the various modules. These aim to provide a stimulus to the critical appraisal of all aspects of the system (process, project or product) rather than lay down specific requirements. Many of the 'check-list' questions are general in nature and the assessor must interpret them as seems most appropriate. Use of these lists depends critically on the expertise and judgement of the assessor. This emphasizes the need for various parts of a company to work together to obtain the best available information.

The model provides a starting point for mathematical modelling of the investment appraisal. Each parameter is given an expected value and the sum of the discounted expected values gives the single, monetary value of the project outcome that is often required in companies for an investment appraisal process. The EMV method provides a basis for calculating a series of outcomes that have a probability distribution, which helps in the use of computer-based risk modelling applications by offering the necessary intellectual framework.

2.7.6 WHO CARRIES OUT THE INVESTMENT ANALYSIS?

Typically, engineering or finance staff have carried out investment analysis in companies. Here another approach is suggested which includes staff in other areas. The potential roles are:

R&D staff
The task of R&D staff is to evaluate the future cost of R&D required due to the investment under study. They also know the risks involved in R&D and can provide this assessment. If patents are involved, the R&D staff may estimate how long it takes to get one and what useful life a patent will have.

Engineering staff
Engineering staff are able to estimate the costs of product and process design and the changes needed. They are also used to estimating the capital and operating cost of a plant or equipment. If an analysis of the potential emissions from a process or product is to be conducted, the engineering staff can best do this. They also know the risks associated with a new process or product. Often they will collect information from different parts of the company and from outside specialists and incorporate it into the investment analysis.

Finance staff
Finance staff are able to help with establishing the cost of finance for the company that is reflected in the discounting factor. They also know the cost of various risk cover tools, such as futures and options. Their help is invaluable in

49

'unbundling' overhead categories into costs that can be directly attributed to projects. They usually negotiate terms and conditions of financing and even insurances. Therefore they need to be informed about the environmental performance of the company.

Marketing staff
Marketing staff carry out the various marketing-related tasks of assessing markets and consumer behaviour. They also estimate the risk and uncertainty of various sales and price scenarios.

Company lawyers
Company lawyers monitor changes in legislation including environmental legislation. They may therefore give guidance on the future direction of environmental regulation and legislation.

Public relations staff
Public relations staff, along with company senior management, are in charge of promoting company image. It is their task to ensure that any investment a company makes in promoting a better environmental image is correctly sized and targeted. The outcome of such promotion must be monitored.

Outside specialists
Depending on the task, a company may or may not have adequate in-house specialists to perform it. In many cases, industrial organizations can provide valuable information on issues that influence the investment decision. Most often such organizations lobby on behalf of the industry they represent.

These roles represent stereotypes that do not have to apply to every company. The main point is that a single department should not try to perform all tasks and generate all information needed for an investment analysis. A focal point for investment analysis may depend on the area to which the investment is directed.

3. EXISTING MODELS AND APPROACHES TO INPUTS INTO INVESTMENT ANALYSIS

3.1 TRADITIONAL INVESTMENT ANALYSIS

There are a number of traditional analysis techniques that provide a technical basis for computing the profitability of an investment in product development, an R&D project or a process technology. These are:

- net present value calculations (NPV);
- internal rate of return calculations (IRR);
- payback time calculations;
- combinations of these techniques.

These models all build on using cash flows, although in some cases asset values and accounting figures are used instead. All of them take into account the time value of money through a discounting procedure. As mathematical models they are straightforward and easy to use. They provide easily understandable decision rules and are deterministic in their nature, satisfying a common desire to have one monetary figure as a means of comparison and an aid to decision.

There are disadvantages in using these techniques when evaluating investments, whether the investments have an environmental impact or not. The techniques provide no guidance as to which cash flows to include. Usually their use is restricted to a functional area of a company. Little or inadequate attention is paid to the various broader aspects of the investment, such as impact on overall company risk, environmental performance or marketability of a product.

The major problem is that the traditional techniques do not allow parameter-specific risk to be included in the equation. This has led to misuse of the models by using various contingencies and abnormally high discount rates. The problems with these techniques are discussed in Section 2.7.2, page 45. The recommended technique is to use expected monetary values (EMVs) of projects. This technique allows risk inclusion, since each parameter is assigned a probability.

Many companies use other evaluation criteria based on financial accounting definitions. The most common one is return on net assets employed (RONA) or return on investment (ROI). A word of caution. The return is defined as financial accounting return and not as net cash inflow. Although eventually these two figures should be the same, the accrual accounting principle of recognizing transactions before the respective cash flow has taken place causes some timing problems.

The asset base may also cause problems. Usually the depreciated value of the assets is used and therefore the return shown is high. Yet the RONA or ROI figure does not indicate whether the underlying assets are being used efficiently. These measures should be used only when the return from an investment is relatively stable from year to year. The picture they give of the profitability of a project is very crude. A further discussion on the comparison of cash flow-based and accounting-based profitability can be found in the Brealey-Myers text *Principles of Corporate Finance*[12].

3.2 TOTAL COST ASSESSMENT

In order to remedy the inadequacies of traditional investment analysis, a total cost assessment (TCA) method was commissioned by the US EPA and developed by the Tellus Institute. The main criticism of the developers of TCA was the narrow selection of costs and returns in the traditional appraisal of projects. They also claimed that 'environmentally-friendly' investments suffered unduly when too short a time horizon was applied in the analysis.

TCA builds on the technique of NPV[13]. Costs are categorized into four levels:
- costs that are normally included and are certain;
- hidden costs, such as monitoring expenses, reporting;
- future liability costs, such as remedial costs and personal injury compensation;
- less tangible costs, such as consumer response, corporate image.

The advantages of TCA over traditional investment appraisal techniques are:
- an enlarged cost inventory. The Tellus studies give guidance on what areas to examine to find relevant costs;
- the model is built on the well-known technique of NPV and it allows inclusion of risk by assigning probabilities to uncertain parameters;
- the model provides a communication mechanism that is familiar in different parts of an organization and all the various parameters from different functional areas can be included if desired;
- due to an enlarged inventory of costs and revenues included in the appraisal, and the consideration of a long time horizon, environmentally-friendly projects gain acceptability through better financial performance;
- the Tellus studies give guidance on how to evaluate sources for liability costs;

- there are three different PC-based software packages that build on the model[6].

The TCA model clearly provides a good starting point for further development in capital appraisal techniques that encompass environmental aspects in addition to the traditional aspects considered. But there are still some problems in the TCA model:

- there is as yet no indication of how to estimate the probabilities of some uncertain parameters;

- conceptually the model focuses on technology and pollution prevention investments, although the technique as such is suitable for evaluating R&D and product design projects;

- the model as yet does not include R&D costs, product design costs or additional marketing costs required when a process or technology is adopted. Technically, the model can be used for these purposes;

- the model pays little attention to the effects on corporate image, and hence on insurance premiums and financing cost through improved environmental performance. Technically there is no hindrance to the inclusion of these aspects.

TCA has been tested in many types of companies. A large study has been conducted within the paper and pulp industry in the US. The conclusions are not decisive, but it is clear that the model is not widely used despite numerous sales of the various software packages. The main problem in companies under study for adopting the model has been the reluctance to make the initial investment in learning the model and searching for the required information.

3.3   LIFE CYCLE ANALYSIS AND ENVIRONMENTAL PRIORITY STRATEGIES

The most commonly used methodology for evaluating the impact of a product on the environment is the so-called life cycle analysis (LCA). This methodology builds on an analysis of a product's life from cradle to grave, trying to estimate the impact on the environment. The impact is estimated from exploitation of raw materials to transporting them to the production site and then, after production and sales, the disposal of the product. The aim is to evaluate *all* effects on the environment and the approach is societal.

The LCA process has four stages:

- initiation;
- inventory;
- impact assessment;
- improvement.

In the initiation phase the goal of the study is defined, based on what decisions should be taken on the results of the LCA. A scoping process takes place when the system, functional unit and data requirements are defined. The problem of functional units needs to be solved. Functional unit refers to the unit of product for which the LCA is conducted — for example, the number of square metres the paint covers, the number of weeks the hairspray lasts in normal use.

The inventory phase provides a detailed account of inputs (such as raw materials and energy) into the system, and outputs from the system. Problems arise in the availability of high-quality data and the transparency of study results. Production and consumption take place in a variety of countries with differing natural environment and infrastructure which contributes to the way environment is burdened. For instance, in many places only one source of external energy is available and so optimization is impossible.

In the impact assessment phase, the full impact of a product on the environment is assessed. Problems arise in trying to value the gravity of the impact and in aggregating impacts into one index. How is a product to be valued that contributes little to global warming but releases a lot of mercury into the environment when used? When publishing the results of this study a peer review of the whole process is called for.

Finally, the results of the LCA can be used to improve the environmental impact of the product. This phase contributes to the sustainable development of society and therefore is a very controversial area. Figure 3.1 shows a simplified LCA for the humble pencil. Further insight into LCA is provided by *The LCA Sourcebook*.

There are several problems in the implementation and use of LCA:

(1) The methodology of LCA is not well established. This gives room for misuse and misinterpretation. It has been noted[14] that company-sponsored LCA studies tend to favour materials and products of the sponsors. A positive exception is the Norsk Hydro PVC study[15]. The lack of an established methodology has led to great variation in studies, even on the same products[11].

(2) Because the methodology is not established, there are misunderstandings of the objectives of LCA studies. The three stages of resource identification, quantification and evaluation have not been fully developed. Often LCA studies finish with identification and no evaluation is done. The major problem in evaluating the actual impact on the environment is that no standard and widely accepted aggregation method has been developed.

(3) As the systems studied using LCA have been very complex, the data requirements have been enormous and in general have not been satisfactorily met. Even well-executed LCAs have at best provided ambiguous results. It is also

54

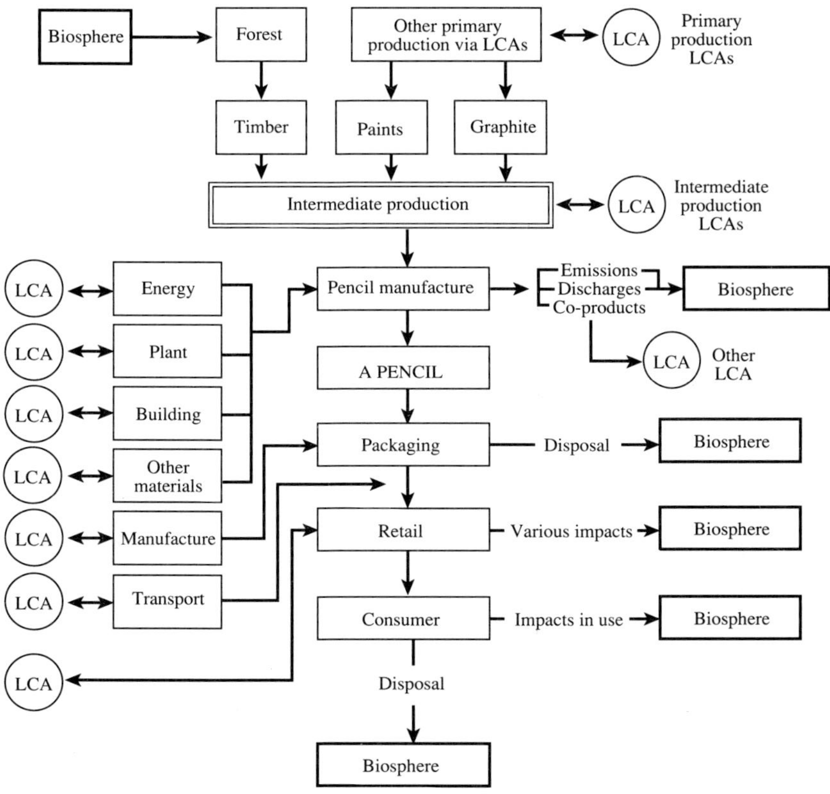

Figure 3.1   A simplified LCA for a pencil. (Source: Gray *et al*, 1993, Accounting for the environment, in *LCA Sourcebook* published by Sustainability, SPOLD and Business in Environment.)

difficult to determine how to allocate resources to products at multi-product sites.

(4)   LCA studies have been used mostly to identify uses of nature's resources. At best, an evaluation of the levels of uses has been made. Generally the total impact of the product or process under consideration has not been evaluated. The LCA studies have been complex, trying to cover all aspects of a life cycle. No attention has been paid to the company perspective — for example, what emissions are relevant to a company and what resource uses are relevant to a company. No effort has been made to assign monetary values to the resource use levels identified.

Although the LCA methodology has its pitfalls and problems, it is still useful. A company must identify clearly what it is trying to achieve with an LCA study. It is possible to use LCA for studying emission levels that are associated with a product or process if the study is restricted to those aspects that are relevant to a company — that is, those that cause a cash outflow or inflow for the company. Therefore the energy used for production is relevant but not the energy used to produce a raw material, such as a particular fat in food production or a solvent in chemical production. Similarly, only the emissions that a company has to pay for, to dispose of, or to discharge into the environment, are relevant. The cost of disposing of the end-product is usually borne by the consumer. LCA can help a company to identify resource uses and to identify which ones are relevant.

As LCA methodology is inadequate in its present state of development, alternative approaches have been developed. One of the weaknesses of LCA is that it is usually a method to identify and describe raw material and energy flow during a product's entire life, where the actual impact on the environment is not valued or estimated. The so-called Environmental Priority Strategies (EPS) system has been developed by the Swedish Environmental Research Institute[16].

This system, based on a simple computer model but a powerful conceptual model, enables product and process developers to evaluate the actual impact of a product or process on nature. The calculations are done by measuring, for instance, how much land is needed for disposal of an emission type. Problems arise when new or unknown compounds are discharged, where no prior information is available on natural resource usage. Additional difficulties arise in deciding what are acceptable emission levels or what the value of clean air is. These subjective judgements depend on the evaluator as well as on society and its values.

3.4   ECO-LABELLING

There are various schemes in different countries to label a product as environmentally acceptable. These eco-labels are intended to assist the consumer to purchase products that are assessed by an independent evaluator to be environmentally acceptable. The basis for these schemes and the premise for awarding a product with such a label varies greatly from country to country. Most of the schemes build on some application of LCA, but often LCA results provide only one of the many conditions for obtaining an eco-label.

The success and popularity of eco-labelling schemes varies greatly from country to country. In Japan eco-labels are relatively easy to obtain and many products already carry them. Japanese producers are of the opinion that an eco-label has improved their sales. In France it is relatively difficult to obtain an eco-label as LCA is implemented in great detail. Hence in 1992 no French product had yet passed the eco-label criteria[17,18].

There are also initiatives to introduce regional rather than national eco-labels. The Nordic countries have established a Nordic label. The EU has set up an EU labelling scheme. (The Nordic and EU eco-label logos are shown in Figure 3.2.) The EU scheme has been delayed and varying levels of interest have been apparent within the community. Some of the community countries have not yet decided on their approach to an EU label.

The labelling schemes present the producer with some problems, for example:

(1) Rules for products vary from country to country. In some cases it can be claimed that eco-labelling is almost a form of trade barrier. It is difficult for a producer exporting to many countries to decide which eco-labelling scheme to apply. It is also problematic that only a few transnational schemes exist.

(2) Rules for the schemes are often applied haphazardly. The juries awarding labels often are not governmental bodies but other interest organizations which use their own discretion when applying the rules. It is therefore difficult for companies to direct their product development efforts as the goal posts are not clearly set.

(3) Eco-labelling schemes have had varying success in different countries and are more important in some countries than others. A company producing a product for various markets may find it difficult to estimate the impact on demand

Figure 3.2   The Nordic and EC eco-labels. (Reproduced by permission.)

of having the product eco-labelled. In many countries there is no data on the real impact of eco-labelling on the sales of a product. Process industry products would seldom directly carry an eco-label, and therefore co-operation with end-product producers is important when estimating the sales impact.

(4) Applying for an eco-label is expensive in time and costs. In most cases, purchasing the right or permission to carry the label is relatively inexpensive. If the outcome of the process is very uncertain and the impact on sales unclear, it is questionable whether a label is desirable.

Eco-labelling schemes can provide some guidance in product design and production process development. An eco-label may create a competitive edge for a product and enhance the environmental image of a company. Before eco-labelling is set as a target in forming a product portfolio, it is important to study the demand for 'green' products in a particular market. In addition, the willingness of consumers to pay a premium for a 'green' product must be established.

At each stage of the investment evaluation process, the evaluator needs to be clear whether the investment will have any impact on the possibility to eco-label the end-product or the production process, as and when such schemes become more common. If this goal is set for the investment being evaluated, the full cost and benefit implications need to be considered in the whole company. The financial benefit from an eco-label on the product can be tested by using conjoint analysis in which consumers are asked to compare the eco-labelling aspect of the product against other product features.

3.5   ECO-BALANCES AND ENVIRONMENTAL INFORMATION SYSTEMS

Process industry companies are becoming increasingly concerned about their environmental performance image, as both authorities and other interest groups require ever better environmental performance from companies. Authorities have determined the way they measure the environmental performance of a company, but other interest groups receive information from companies in a much more varied fashion; there are no rules or legislation in place in the Western industrialized world on how environmental performance is to be measured objectively.

Eco-balances and energy and raw material balances have been partially developed to enable companies to give information about their environmental performance and also for other interest groups to form a picture of their

Inputs, kg		Outputs, kg	
Purchased and in-house scrap	1049	Products	1000
Limestone	9	Carbon dioxide	19
Lime	30	Carbon monoxide	12
Flux	5	Other gases	21
Oxygen	11	Waterborne waste	1
Iron ore	5	Slag	105
Carbon electrodes	5		
Air	100		
Moisture	4		
TOTAL	1218	TOTAL	1218

Figure 3.3   A simple example of a mass balance. (Source: Franklin Associates, as appeared in *LCA Sourcebook* published by Sustainability, SPOLD and Business in the Environment.)

performance. Figure 3.3 gives a simple example of a mass balance. Increasingly companies are building systems to monitor their own performance in order to identify areas of concern.

These internal information systems are sometimes called environmental management systems and mostly cover technical data. Financial data on costs of environmental management and liability potential, and the insurance cover for it, have not usually been incorporated into them.

Internal environmental information systems are necessary when a company is trying to estimate the financial consequences of processes and products. External information media are important for external evaluators of the environmental risks of a company. The most common external information system takes the form of an eco-balance, although many companies publish verbal descriptions of their environmental protection and improvement activities, along with some numerical information on emission levels.

Eco-balances can take several forms[19]:

• they can be social balances that describe the interfaces between a company and its environment. Inputs and outputs are evaluated in monetary terms and the social impact is assessed. The target group for the information is employees and the general public;

- they can be in the form of ecological bookkeeping and the use of natural resources is aggregated into one figure. The target is to express the environmental burden and the use of natural resources in monetary terms;
- they can be a set of environmental ratios and indicators. The goal is to express the use of resources from the point of view of utility created for society;
- they can be a balance of energy and raw material use. The impact on nature is estimated using scientific methods;
- they can also be built on an LCA of a product or a process in which the total energy and raw material usage is identified, and some qualitative statements are made of the impact on nature.

All these descriptions suffer from a non-established methodology, which means that they are not restricted to a clearly defined entity, the aggregation problems are not solved and valuation problems still exist. This means that they are not comparable across companies, products or processes. They are also open to misinterpretation and misuse.

Although companies wish to use external information systems to enhance the perception of their environmental performance in the eyes of the stakeholders, there is no one uniform and established way of doing it. It is therefore difficult to measure the impact of improvement in the environmental performance on the company, particularly the monetary impact. The only way to measure the impact is to conduct enquiries of the market study type amongst chosen interest groups.

The insurance industry is in the process of developing products that it can offer to companies with potential environmental liability problems. Both in the US and in the UK the industry is prepared to insure companies against sudden accidental risks. Gradual pollution liability cover is still under study in both countries[20,21,22]. The banking industry is already formulating policies for environmental audits to be conducted at sites of potential credit customers, and specifically targeted 'green loans' are available for environmentally acceptable.

It should be noted that the measurement of environmental performance, potential environmental risk and future liability costs are very much under development. Also legislation concerning publication of information about environmental performance and risks is being prepared in many countries, but to date no clear rules exist.

Therefore the financial impact of a good environmental image is difficult to evaluate.

3.6 OTHER APPROACHES

Assigning a monetary or a subjective value to emissions is much researched in academia. The Institute for Applied Environmental Economics (TME) in The Netherlands has developed a 'Decision Model for Environmental Strategies of Corporations' (DESC)[23]. This model is a computerized tool for evaluating the best waste management route for different products and processes. The program offers more than a thousand solutions with the associated cost of emission reduction. The model was initially developed for Unilever. The impact on nature is evaluated from a societal point of view, but can offer a valuable contribution when estimating future liability costs.

The Finnish Water and Environmental Agency has developed a model for calculating the negative impact of a production site on air and waterways[24]. The perspective is societal and many judgements on the degree of negative impact on the environment are subjective. The approach is by no means fully scientific and objective yet the results of the study have gained much publicity. The approach can, however, provide companies with valuable information on the potential of future liability problems.

The number of different approaches to evaluating the impact of a process or a product on the environment highlights the importance of these questions to society, its different interest groups and companies making efforts to clean up their production. The approaches have a societal perspective, but can offer significant guidance to companies by identifying potential sources of liability issues. The techniques themselves can also be used by companies, if the emissions or waste treatment aspects relevant for the company are evaluated. These aspects are fewer than the ones of interest to society, as companies do not always pay for all the pollution they cause.

3.7 SUMMARY OF PRESENT APPROACHES AND THEIR INADEQUACIES

Although many of the present approaches can serve as valuable building blocks when evaluating the financial consequences of environmental investments, be they investment in R&D, product design, process changes or product or company image, they all still have many inadequacies:
- the present models cover only part of the relevant parameters and many of these are incorrectly included. In many cases relevant sub-categories are completely omitted;
- the present models do not provide a holistic conceptual model that covers all the necessary aspects of decision-making around environmental investments;

- there is great concentration on issues related to technology, but issues such as marketability and corporate image are little studied or ignored;
- parameters that are deemed difficult to estimate are left out. This is particularly true with future liability costs;
- many of the models have a societal rather than a company perspective when evaluating the environmental impact and its monetary value;
- most of the models do not include uncertainty or the inclusion is done inadequately.

4.　TEST CASE A — PILOT PLANT FOR NON-CHLORINE BLEACHING OF PULP

4.1　INTRODUCTION

4.1.1　THE COMPANY

The company undertaking this investment is Wisaforest Oy Ab, a subsidiary of the Finnish Kymmene Corporation that was formed from three major paper and pulp producers about three years ago. The Kymmene Corporation subsidiaries are old companies, most with at least a hundred year history in paper and pulp making. Kymmene Corporation has operations in five European countries.

Wisaforest Oy Ab turnover in 1992 was FIM1851 million which was around 13% of the corporate turnover. Wisaforest production is dominated by pulp, kraft paper and sawn timber. The main production site is in Pietarsaari on the north-western coast of Finland. Approximately 60% of the pulp production is sold internally, mainly to the German paper mill Nordland Papier. The rest is used in Wisaforest's own paper production or sold to outside customers.

The Wisaforest Oy Ab production site at Pietarsaari, Finland

4.1.2 HISTORY OF THE INVESTMENT UNDER STUDY

The main method of making chemical pulp today is kraft pulping. The fibres released from wood chips by kraft pulping are dark and require bleaching before the pulp made from them can be used in manufacturing of communication quality paper. Traditionally, kraft pulp was bleached with chemicals containing chlorine. The most commonly used chemicals were chlorine gas or chlorine dioxide solution. Previously, hypochlorite solution was also used widely.

It has long been established that waste waters from the bleaching of kraft pulp with chlorine gas produce emissions that can have a mutagenic effect — for example, in water fleas and larvae. Untreated waste waters from conventional bleaching of kraft pulp are toxic to the receiving waters. The paper industry has reduced the use of chlorine gas over the years and substituted it with chlorine dioxide and other bleaching chemicals — for example, oxygen delignification. Treatment of waste water in mechanical clarifiers and activated sludge secondary treatment facilities has greatly reduced the harmful effects in the receiving waters.

In 1985, significant quantities of dioxin were found in river sediments below pulp mills in Sweden and the USA. At that time, there was great concern about the possible carcinogenic effects of dioxin due to the use of this chemical in the Vietnam war, and due to the Seveso accident in Italy. Various 'green' interest groups started accusing the paper industry of producing highly dangerous environmental pollution due to the use of chlorine in pulp bleaching. The American Paper and Pulp Association was able to prove that the paper and pulp industry was only a minor contributor to the level of dioxin in nature, but public interest had been awakened.

In Germany, Greenpeace took the issue of chlorine seriously and started to demand totally chlorine-free pulp even though dioxin occurs naturally in peat in the Scandinavian marshland, and dioxin produced in pulp bleaching is only a fraction of the dioxin produced by other industrial or municipal activities. It is worth noting that the German pulp industry produces only sulphite pulp that can be bleached easily without chlorine chemicals — for example, with oxygen and peroxide. Paper produced from sulphite pulp is somewhat weaker than paper based on kraft pulp.

As the concern over the dioxin threat in pulp bleaching and in paper products started disappearing in 1990, Greenpeace retargeted its campaigning at chlorinated organic materials in the waste waters of pulp bleaching facilities. The generic name for such chlorinated organic materials in aqueous effluent is AOX, or absorbable organic halogens. Greenpeace and other environmental interest groups demanded zero AOX levels in the waste waters of the pulp and paper industry. This can only be achieved by totally discarding the use of any chemicals containing chlorine in the bleaching process.

Although only the German-speaking part of the European market was demanding totally chlorine-free (TCF) pulp, its importance as a market was so great that companies in Europe started to find alternatives to chlorine bleaching. In Sweden chlorine was replaced by oxygen and peroxide. The brightness achieved was measured at 78–80%. This was considered adequate in Germany, but the paper produced from this pulp took on a slightly yellow tinge when exposed to sunlight or fluorescent lighting.

It has long been established that ozone gas has a bleaching effect. The major hindrance to using it in pulp bleaching has been the detrimental effect on the strength of pulp fibres. The use of ozone as a bleaching agent was tested in laboratory conditions in several companies. The quality of paper produced was better than that provided using oxygen and peroxide bleached pulp. The brightness could be measured at 85–86%, there was less tendency towards a yellow tinge and the paper was of acceptable strength. Pulp can also be bleached by using a combination of oxygen or ozone, peroxide and some enzymes.

While Wisaforest was carrying out bleaching experiments at laboratory scale, Union Camp announced its decision to build a full-scale bleaching plant using oxygen, ozone, peroxide and chlorine dioxide. Although this method is not totally chlorine-free, it was claimed that using chlorine dioxide at the last bleaching stage would not release any significant amounts of chlorinated organic materials into the waste waters of the bleaching plant.

4.1.3   OPTIONS FOR WISAFOREST

During 1990 Wisaforest and Kymmene Corporation had to develop their strategy with respect to totally chlorine-free (TCF) pulp. The options were:

* purchase it from outside suppliers;
* develop a technology for their own production;
* wait for others to develop the technology and buy this technology when it was ready. In the meantime, Kymmene Corporation would have to buy in TCF pulp from outside and sell all its own standard pulp production outside German-speaking markets;
* start producing pulp using chlorine dioxide as the chlorine-containing chemical, and address non-German-speaking markets with this pulp. (This is called elementary chlorine-free (ECF) pulp. Research has already shown that waste waters from ECF production have only a minor amount of AOX material.)

When Kymmene Corporation was formed, one of the main strategic cornerstones was the achievement of good self-sufficiency in pulp production. Although the Corporation is a small net buyer of pulp due to some special grades, it has achieved a good level of self-sufficiency. The Wisaforest factory was producing mainly for the German market and moving away from that market was not feasible: the marketing channels presented a bottleneck. (Finnish pulp producers sell most of their pulp through a common marketing organization, Finncell.) The strong Finnish currency made Finnish pulp too expensive to be exported to distant countries, such as Japan. (This situation has now changed due to two devaluations of the Finnish currency.)

The only feasible alternative, as perceived by Kymmene Corporation, in the strategic and market conditions of 1990, was to develop its own TCF pulp production. The decision was consistent with the overall strategy of Kymmene. The Corporation therefore studied the potential markets and price levels for TCF pulp. Initially, there was a price premium on TCF pulp, but this was expected to disappear by the year 2000.

Ozone bleaching technology was tested and performed well in laboratory conditions. It was, however, not tested on a larger scale and there remained many technical unknowns in the process. Some of the technical risks were to be borne by the co-operation partners chosen for this project and, partially, by a technology investment fund (TEKES). The fund required money to be paid back only if the project was successful. Estimates were made on the potential variable production costs of TCF pulp per ton and the extra margin gained from selling TCF pulp when standard pulp demand was expected to be falling. On this basis, it was possible to gain an idea of the financial impact of the production of TCF pulp in a pilot plant.

It was decided that a commercial-scale pilot reactor for ozone bleaching should be built. The capacity of the pilot plant was to be such that it could process a quarter of the pulp production of the Pietarsaari site. The pilot reactor was also designed to accommodate experiments with both hardwood and softwood pulp. Agreements were drawn up with the chosen project partners on exploitation of the developed technology and with a potential fall-back situation if the experiment were to fail. The pilot ozone reactor was placed at the hardwood pulp line at the Wisaforest site.

4.1.4   THE INVESTMENT DECISION PROCESS AT KYMMENE

Kymmene Corporation is organized into several independent subsidiaries which have their own technical, marketing, financial and general management staff. The investment decisions are co-ordinated by a corporate investment department which ensures that investments are in line with the overall long-term company strategy and, in particular, with the technology strategy.

The Corporation has not set specific rules for the acceptance of investments, such as hurdle rates. There are, however, strict rules on how an investment proposal should be presented, but no rules for specific time horizons for

Unit process — ozone bleaching

appraisals exist. Usually an internal rate of return (IRR) is calculated. All market forecasts are made or commissioned by the operating units. No specific rules exist for incorporating environmental aspects into investment appraisals. The company policy is to improve the environmental performance of the company and use advanced environmentally-sound technology when its economic performance is known. Hence the major portion of investments made in Wisaforest in 1992 was directly targeted at improving the company's environmental performance.

4.2   TEST CASE DESCRIPTION

4.2.1   THE INVESTMENT UNDER STUDY AND THE ADOPTED APPROACH

Kymmene Corporation had requested a review of its investment decision made in 1990. As the investment in TCF technology was regarded as highly strategic and absolutely necessary for Kymmene, the original numerical investment analysis was less detailed than might otherwise have been the case. An example of the financial impact of the investment was calculated for the technology fund supporting the investment; this was based on a prediction of Kymmene Corporation and Wisaforest concerning market demand and the expected price for TCF pulp.

The particular investment that is being analysed using the model developed in this study is the pilot reactor for bleaching hardwood pulp. Although the pilot plant is suitable for softwood bleaching experiments, the bleaching capacity of the plant was designed for hardwood pulp production at the Pietarsaari site. Wisaforest did not conduct any numerical analysis of its alternatives and hence these are only broadly addressed in this test case.

Wisaforest produced the numerical material required by the model. There were some difficulties in providing all of it. This material was different from that which they would normally use in their analysis in:
- the treatment of overhead and fixed costs;
- the definition of the project.

The production processes and site structure are fairly stable in Pietarsaari and therefore division into variable and fixed costs is valid in day-to-day cost monitoring. In a case where a unit process or site facility — such as a power station or waste treatment facility — is altered, division into fixed and variable cost poses some problems. The cost pools have to be broken into their components in order to take into account the change in costs due to a new cost element. The treatment of overheads is traditional at Wisaforest and therefore hides some important cost elements. This is particularly true for marketing costs.

The pilot reactor project was originally analysed solely as a separate R&D project disregarding its implications on the overall plant if the results of the R&D project were implemented in production. Yet the income from the project was calculated for the total technical lifetime of the project. The full respective investment, which included changes in production lines and in facilities common to all plant, was ignored.

This study offers an alternative way of evaluating the investment. The test case will highlight the differences between a traditional approach to investment appraisals and the model. The two main areas of emphasis are:
- demonstrating that an R&D project is not an isolated project, but needs to be considered in the context of the whole plant, including its implications on company income;
- incorporating risk and uncertainty into the evaluation in a structured fashion.

Wisaforest is keen to gain experience in dealing with risk. It is also important to highlight the difficulties in locating adequate information and to make some suggestions concerning the approach to acquiring information. All values in the examples are disguised — even the relative size. Only the comparison between the two approaches is given as actual numbers.

68

The Wisaforest approach is tied to company strategy and is multidisciplinary in the evaluation of a complex situation. Wisaforest and Kymmene Corporation bought themselves a 'technology option'. Today they are in a position to choose whether to go ahead with ozone bleaching, and expand it to cover all Wisaforest pulp bleaching. Had the pilot plant not been built, this option would not exist. In buying such an option a company may well accept a short-term loss and negative net cash outflow, if the prospect of the technology is deemed positive in the long run. Wisaforest did exactly that. It partially analysed the investment in the technology, and then estimated what it would mean to the company in the long run, both in financial returns and in strategic terms.

4.2.2   TRADITIONAL VERSUS MODEL APPROACH

GENERAL ASPECTS

Project definition
Wisaforest had originally defined the project solely as R&D. Yet in its actual calculations the company included revenue items which it was possible to achieve only with some additional investments, such as changes in the relevant production lines and expansion of the pulp cooking facility. The lack of cooking capacity causes a bottleneck in production. This has a significant impact on the whole investment, both from a financial and technical point of view, and should therefore be included. The additional investment required to expand the cooking capacity for ozone bleaching is larger than the total investment for building the bleaching process. There were also some other changes in the production line required due to TCF production. These represent an additional 12% of the initial investment cost.

Calculation basis and fixed versus variable costs
Wisaforest carried out its original calculations using an incremental basis. This is appropriate as the investment under study represents a change in existing circumstances. This approach is also chosen for the new calculations. The main problem in Wisaforest's initial calculations is the use of fixed and variable costs. Although, in their normal day-to-day cost monitoring, this division is sensible, in this case it distorts the picture. Wisaforest had assumed that its fixed cost base would remain unchanged. This may be the case, but its fixed cost definition includes plant insurance. A new unit process may change the safety classification of the plant and hence the insurance cost.

Time horizon

The technical age of the plant was estimated at 15 years. All original cash flow calculations were related to this time frame. There is no reason to expand the time horizon as no gradual environmental harm is expected. The process emissions are well identified and analysed by Wisaforest. It had also discussed the new process and its emissions with the environmental authorities. All operations were conducted according to existing legislation. There were no expectations of radical changes in the environmental legislation. An environmental impact analysis is in progress to gain a full understanding of the effect of 100% ozone bleaching on the ecology of the receiving waters. Preliminary results of the study will be available when the decision on capacity expansion is made.

Discount rate

In this case Wisaforest did not use traditional discounting techniques but calculated both an IRR and a payback time. Risk is not included in its calculation in a particular way. Normally, the company conducts a sensitivity analysis, altering the prices of various key production inputs such as raw materials.

Originally no such analysis was conducted for this case. The expected selling prices of TCF pulp were assessed by the local management team on the basis of in-house research and discussions with market experts. The estimates of the studies, which coincided well, were used in the original calculations. For the model calculations, new prices were estimated at different levels of certainty.

Overheads

Overheads proved to be a difficult area: Wisaforest attributes overheads to projects on the basis of an overhead percentage. In this particular case it leads to some distortions. Waste management will become less costly due to a change in the character of the aqueous effluent from the bleaching phase in production. Overheads allocated according to a non-project-dependent percentage will understate the effect of cost reduction.

Depreciation

Wisaforest has indicated that it normally includes depreciation in its cash flow calculations. But depreciation can be included only when calculating the tax effect of an investment. Once that has been discovered, depreciation should be added back into the cash flows, as it is not a cash flow itself.

When Wisaforest made a new set of calculations to calculate IRR and the payback period, it did not include depreciation. This is due to the way the company defines its costs. It used only the fixed and variable costs to do the calculations and these do not include depreciation. Thus there was a slight distortion in the numbers.

Interest charges and project financing cost

In normal circumstances Wisaforest attributes an interest charge to a project on the basis of working capital. In this case, these charges were considered unchanged and hence excluded. But Wisaforest should analyse this situation, as the time from one production phase to the other changes. The change is due to the longer cooking time the pulp requires when ozone is used for bleaching. This delay initially requires some extra time in interim storage before the pulp is admitted to cooking.

Due to a change in ownership that took place within Kymmene Corporation some three years ago, Wisaforest is burdened with some interest payments to the parent company. These interest costs are not to be included as a cash outflow in the project appraisal. The free net cash flows from the project are to contribute to the payment of these interest costs. If the interest payments are included as a cash outflow in the project, the project is expected to cover the interest payment twice.

Wisaforest is also charged a corporate interest cost that is believed to be calculated on the basis of the existing fixed assets that are attributed to Wisaforest. These assets create an opportunity cost to the Corporation and Wisaforest at the corporate weighted average cost of capital. This opportunity cost is the cost of the fact that Kymmene is investing its funds in fixed assets rather than in other forms of investment vehicles, such as shares or bonds. The investment in the fixed assets of this project is to produce an income that represents the return on this specific investment. This investment is not to yield a return on previous investments. However, the cost of any changes that are required in other assets is to be included in the investment appraisal.

A common mistake is to regard cash in the bank account as 'free' of charge — that is, that no interest is paid for this cash. But although no interest needs to be paid to the bank for using this cash for an investment, there is an opportunity cost for this cash. This is the interest that is lost because the cash is used in the investment instead of other purposes. It is to be remembered that the interest charges paid to a bank are not to be included as a cash outflow. The investment made using borrowed cash is to produce a cash inflow large enough that the bank charges can be covered.

Wisaforest received a grant and a subsidized loan from a technology fund. The grant represents an income that certainly accrues to Wisaforest and should be treated as such. The loan needs to be paid back only if the project succeeds. The difference between a subsidized interest rate and the company's normal financing cost should be treated as an income to the project. Originally, Wisaforest was not certain whether the grant and loan were to be granted, and hence these sums were not included in its original calculations.

RESEARCH AND DEVELOPMENT

Costs

In the original calculations, the costs for R&D were estimated according to the appropriate categories. No particular overheads were allocated to R&D; management of this work was included in the overall management overhead. The largest omission, due to the TEKES technology fund rules, was in the area of testing costs. Equipment and labour to be used for testing were estimated; the material to be used for testing was not. It was expected that pulp bleached using ozone could be sold at the price of standard pulp or at a slight discount. The cost impact of this discount was not calculated originally but is now included.

A Hazop analysis was conducted of the new unit process. On this basis, a probability was assigned to an accident occurring. This was not costed into the project. On the basis of the company risk management policy, a monetary value is now calculated for this risk and included in the model calculation.

Revenues

The project received a special grant from the TEKES technology fund. Originally, this was not included as an income for the project. The interest subsidy of the loan should also have been included as an income for the R&D part of the project as the loan was obtained solely due to the R&D nature of the project. This has been done in the model calculations.

The new unit process meant that more waste wood and timber was available for incineration in the power station of the plant. The extra income from this was correctly included. Regulatory compliance was studied and no change was indicated by the authorities. Hence, correctly, no change was included in the calculation for this factor.

Risks

Two types of risks can be identified in the case of R&D: technical risk and progress-related risk. These were not included in the initial calculations. The technical risk relates to the certainty of reaching the target quality level. This risk estimate was used to evaluate whether the TEKES loan had to be paid back.

Wisaforest estimated the risk of not adhering to the project timetable and the associated additional costs and savings which would accrue if the timetable was not adhered to. These estimates were then used to calculate the expected monetary value (EMV) of the R&D costs of the project in the model calculation.

PRODUCTION

Costs

In the original calculations, the actual *capital investment* in the building of the pilot reactor and the facilities related to the reactor was accurately estimated. The greatest omissions were the changes required in the common facilities of the plant, such as the power station, waste treatment facility and the cooking facility. Some of them were small, but the extension of the cooking capacity was a large investment. In addition, some other changes were required in the production line. These were included in the model calculation.

Measuring and alarm devices for *emission monitoring* were included correctly. Patents and licences were assumed to be a part of the overall cost of equipment. These assumptions were maintained in the model calculation.

The *variable operating costs* were calculated accurately. Water consumption was estimated to decrease somewhat but its cost impact was regarded as small. Waste treatment was simplified. The changes needed were not costed but were estimated to be small. Emission permits did not require changes. No alternations were made to these assumptions in the model calculations.

Plant maintenance, insurances and quality monitoring were included as part of overall overheads. In this particular case these cost items might have been small but they needed to be estimated. As this was a new product, it could be expected that quality monitoring would be costly. This needed to be identified as a separate cost item. There was not adequate information available to include these into the model calculation in other than original format.

Material costs were assumed stable in relation to each other. An assumption of overall cost inflation was made and used for all cost types. In the original and the model analysis, no assumptions were made about the uncertainty surrounding the actual material costs. Wisaforest had originally built in a buffer for price variances. Although this was actually an incorrect way of including uncertainty, as no probability was assumed for this buffer, the buffer was kept intact in the model calculations. In normal cases, Wisaforest conducts a sensitivity analysis of the prices of the most important materials. The drawback of this approach is that no probability is assigned to each scenario.

Financial costs are usually estimated on the basis of the working capital needs. Wisaforest uses an interest rate for working capital which is different from the one used for other asset costs. The reason is not clear, as a company has only one marginal cost of capital or one weighted average cost of capital.

Potentially, Wisaforest considers working capital and other assets to be financed from short-term and long-term financing respectively. These forms of financing usually have a different market price. The correct rate to use is the weighted average cost of capital to the company. A good estimate is to use the marginal cost of capital.

Exchange rate variances or the cost of foreign exchange cover were not included either in the original calculation or in the model calculation. Foreign exchange rates were estimated to remain stable for 15 years, which could be regarded as optimistic. The cost of foreign exchange cover is difficult to calculate accurately when the company covers net positions — that is, the net of outflows and inflows of a currency — rather than the flows of a project. An estimate can be made by calculating the gross cover needed for the project flows.

In this case, the cover represented a potentially large gross revenue item as there was an inherent interest rate gap between the Finnish markka and the foreign currency in which the production was sold, the German mark. An added complication is that pulp is priced in US dollars but sold in German marks. Therefore an estimate should also have been made of the relative exchange rate between the German mark and the US dollar. In the model case, the exchange rate uncertainty was included through a risk factored exchange rate used when calculating sales income in Finnish markkas.

It can be argued that the TCF sales to Germany replace standard (STD) pulp sales to the same country. Therefore the only additional cost for exchange cover comes from covering the price premium. This was assumed to be small in the model case and therefore not included.

Impact on other *overhead* costs, such as procurement, accounting or management, were not included as they were regarded as small. The impact on *direct labour cost* was not estimated separately. Instead, an overall estimate of the change in variable costs was calculated. This change in cost was then used for identifying the increase in operating margin when producing TCF pulp instead of standard pulp. These assumptions were maintained in the model calculation as no estimates for the changes were available.

Revenue

The revenue from TCF pulp was calculated on the basis of an estimate of sales volume and sales prices. The fixed cost basis was assumed stable. The variable cost basis was altered according to the expected changes in variable costs due to the new bleaching method. All elements in the calculation were assumed to remain on the estimated level or follow an increase/decrease pattern as forecast originally by Wisaforest. The increase in operating margin was thus calculated for each year for 15 years into the future. No other cost or revenue items were included. These were either regarded as insignificant, as described previously, or omitted whether by mistake or choice. The original calculations were done on a high level only. The only savings included were the ones from increased energy production as a by-product.

Risks

The largest risks in the overall production area come from the raw material prices, the actual capital cost of the new unit process and the changes required in other facilities. Raw material prices were actually expected to decrease but Wisaforest built in a buffer for raw material price increases. This buffer was considered to be the same in all years under consideration. Project delays or availability of capital were not considered to be sources of risk and hence no buffer was built in for these. In the model case these assumptions were retained as they were in the original calculations due to lack of information.

MARKETING AND MARKETABILITY

Costs

Project-specific marketing costs were treated as part of the overall marketing cost in overheads. The project required specific forecasts for demand and price conditions for several countries. This cost should have been singled out from the overall marketing budget.

Actual sales costs were believed not to change over the lifetime of the product. The same marketing channels were expected to be used. The possibility that new wholesalers might have been approached was considered, but no additional cost was allowed for this. Overall, the distribution cost was regarded as the same for both.

This product would have been potentially eligible for eco-labelling in many countries, as the new bleaching method was developed due to demands from environmentalists. No cost was assigned to obtaining such a label nor revenue enhanced for having the label.

Revenue and related risks

Wisaforest made thorough studies of the market expectations. As these coincided, both within and outside the company, no buffers were built into the demand or price estimates. The actual sales revenue depends on three factors: volume, price and exchange rates. The probabilities of each level of these have been incorporated into the model calculations. As Wisaforest did not consider eco-labelling, no extra price or revenue impact has been estimated for an eco-labelled product.

OTHER MODULES

The emphasis of this analysis is on R&D, production and marketing. Product design and company image have not been analysed. Product design is an integral part of the overall R&D work and hence has not been singled out. Company image was not a monetary consideration for Wisaforest although it has been in the forefront of improving products environmentally. Kymmene Corporation overall wished to improve its environmental performance and this project was regarded as one of the means to achieve that goal. Wisaforest has a long tradition of trying to achieve environmental targets set by authorities and it invests continuously to improve and monitor its environmental performance. Wisaforest has not conducted studies into the financial impact of those improvements.

4.2.3   ALTERNATIVE CALCULATIONS

Revised calculations are demonstrated below. The first section shows the calculation of the new cash flows and other contributing factors are shown in detail to demonstrate mainly how risk can be included in the parameter values. This is followed by an EMV calculation, given in the form of a table. The figures are fictional. The third section makes a comparison between the original and model results.

NEW ANNUAL CASH FLOWS AND OTHER FACTORS

Sales income
(1) Exchange rate
Expected exchange rate was originally 1 USD = 4.3 FIM.

The following probabilities were given by Wisaforest. The probability $(= p)$ of the exchange rate:

(i)   remaining on this level, $p = 0.06$;

(ii) exceeding this level by 10%, $p = 0.94$;

(iii) being lower than this level by 10%, $p = 0$.

The new expected value of the exchange rate in the model calculation is therefore:

$$[0.06 \times (4.3 - 10\%)] + [0.94 \times (4.3 + 10\%)] = 4.70 \text{ FIM to 1 USD}$$

(2) Demand volume

Wisaforest gave its prediction on the probability of the demand level in the following way:

(i)   remaining as predicted, $p = 0.385$;

(ii) being 10% less than predicted, $p = 0.075$;

(iii) being 10% higher than predicted, $p = 0.54$.

Each demand quantity was then multiplied with these probabilities to get to a probability-weighted demand for each year and each market segment. The same probabilities were assumed to apply in each market.

(3) Price

In the original Wisaforest calculations the underlying price used as a guide to the TCF pulp price was the standard (STD) pulp price. This was predicted to increase from a base level at an estimated rate. The TCF pulp price was calculated as a sum of the STD price and a price premium. The original prices were given in FIM and at an assumed stable USD level.

The new prices were calculated by multiplying the price premium with a risk measure. This can be regarded as somewhat dangerous as the world market price for pulp is very volatile. The volatility was not estimated on the basis of time series data, which could have been a feasible alternative and used, for instance, in money market option pricing. The risk-weighting of the premium was thought to be the most interesting parameter.

The risk-weighted price premium was calculated on the basis of Wisaforest's expectations. Asking correctly stated questions about the expectations is important. The human mind seems to think in point probabilities — it is 'easier' to estimate the probability of a price rather than asking the price at a certain probability level. To get to a statistically-correct answer, Wisaforest was asked to estimate:

• the price premium that it felt it would exceed with 90% probability;
• the price premium that it felt it would exceed with 10% probability.

The most expected price premium should fall between these two prices. Due, perhaps, to the remoteness of the actual decision, the price premium predictions were very different from the original expectations. Also, predicting a series of prices can become difficult and a schematic approach is easy to adopt. In the model calculations, Wisaforest's predictions are used despite some difficulties in identifying the precise nature of the prediction in each case.

Tables 4.1 and 4.2 on pages 78–79 demonstrate the sales income calculations. The figures are fictional.

TABLE 4.1
Export sales: disguised figures

	1993	1994	1995	1996
Price of standard pulp, FIM	1000	1500	2000	2500
Price in USD (4.30)	233	349	465	581
Risk-weighted price premium, FIM	500	500	500	250
Premium in USD (4.30)	116	116	116	58
New price in USD	349	465	581	640
New price in FIM (4.7)	1640	2186	2733	3006
Expected sales, Finncell export (softwood TCF pulp)				
Risk-weighted export demand	0.01	0.1	0.3	0.4
Risk-weighted export price	1640	2186	2733	3006
Risk-weighted sales income, FIM	16	219	820	1202

TABLE 4.2
Domestic sales: disguised figures

	1993	1994	1995	1996
Expected price, domestic sales (hardwood)				
Price of standard pulp, FIM	2000	3000	4000	5000
Risk-weighted price premium	500	500	500	250
New price, TCF pulp	2500	3500	4500	5250
Expected domestic sales income (hardwood TCF pulp)				
Risk-weighted domestic demand	0.1	1.5	2.5	3.5
Risk-weighted domestic price	2500	3500	4500	5250
Risk-weighted sales income, FIM	250	5250	11,250	18,375

Grant

In the original calculations the 'free' income from the grant received was ignored. Although it does not represent a large figure, it was included in the calculations in the model.

1997	1998	1999	2000
3000	3500	4000	4500
698	814	930	1047
250	250	150	150
58	58	35	35
756	872	965	1081
3552	4099	4536	5083
0.5	0.6	0.7	0.8
3552	4099	4536	5083
1776	2459	3175	4066

1997	1998	1999	2000
3000	4000	5000	3000
250	250	150	150
3250	4250	5150	3150
4.5	5.5	6.5	7.5
3250	4250	5150	3150
14,625	23,375	33,475	23,625

Loan effect

Wisaforest received a subsidized loan for financing part of the project. The loan was to be paid back only if the project succeeded. The interest differential between the loan interest and Wisaforest's normal interest cost is a real earning from Wisaforest's point of view: without the project Wisaforest would not have received the loan. The interest rate subsidy means that Wisaforest needs to earn less from this project than in the case where the project was financed from normal sources. The earning is calculated on the basis of the interest differential. Wisaforest gets to keep the interest differential, whether the project succeeds or fails. No cost is assigned to having to finance the reimbursement of the loan in case the project succeeds.

Assumptions:
- loan amount   3.2 million markkas (= Mmk);
- borrowing time   7 years;
- grace period   4 years;
- amortization   once a year, at the end of the year;
- amortization   5th year = 1 Mmk,
                      6th year = 1 Mmk,
                      7th year = 1.2 Mmk.

Interest:
- expected Bank of Finland rate   8%;
- expected Kymmene Corporation rate   11%;
- loan rate   Bank of Finland rate − 3%;
- expected loan rate   5%;
- interest differential $11 - 5 = 6\%$.

Testing cost

Testing costs, other than actual test pulp, have been included in the overall investment cost. This cost represents the discount that has to be given for TCF pulp of inferior quality, but sold to the market. The amount used here represents a gross generated for the purposes of this study by Wisaforest and is based partially on real figures.

Future liability cost

A Hazop analysis was carried out on the unit process. It gave an estimate of 0.1% probability of an accident. Wisaforest and Kymmene Corporation have an insurance policy that covers costs for accidents that cause hazards for more than 50 Mmk. Hazards relating to accidents with less than a total of 50 Mmk are covered by the company. On this basis the following crude estimate for the potential value of a future liability is calculated. It is assumed equally likely every year.

$$0.5 \times 50.0 \times 0.001 = 25{,}000 \text{ mk p.a.}$$

The company insurance and risk management policy are based on an unknown probability distribution. Therefore, a binomial distribution is assumed.

Cooking facility

On the basis of the assumptions that:
- the cooking facility is needed to support the expected production volume that would bring the expected sales income;
- Wisaforest was aware of the potential need for debottlenecking TCF pulp production by expanding the cooking capacity;

the cooking facility and debottlenecking investments are included in the overall project appraisal. No probability of the costs is attached, but the procedure would be similar to those described already.

Changes required in the production line

On the basis of the assumption that Wisaforest knew that these changes were required due to TCF pulp production, the investments in the changes are included. No probability is attached to the amounts.

Original project cost with risk included

The assumptions here are that:
- the original budget holds at a probability of 89%;
- the budget is exceeded by 10% at a probability of 5.5%;
- the budget is undercut by 10% at a probability of 5.5%.

The expected total investment budget is therefore:

$$(0.89 \times 84.6) + (0.05 \times 93.1) + (0.05 \times 76.1)$$

= 84.6 Mmk, which is the original project cost. This result is due to the same probability given to the undercutting and exceeding of the budget by 10%. In other words, no new information is gained by this type of risk assessment; the risk of undercutting has to differ from that of exceeding the budget for this calculation to add new information.

Cost of project progress risks

If the project does not stay within its timetable, both production and testing will be delayed. This means that revenue from the project is delayed and even more

TABLE 4.3

Total project EMV calculation (figures are fictional)

	1992	1993	1994	1995	1996	1997	1998
Cash inflows							
New operating margin	0	1000	6000	12,000	20,000	30,000	30,000
Grant	1000						
Interest income, TEKES loan	163	163	163	163	163	112	61
Total cash inflows	1163	1163	6163	12,163	20,163	30,112	30,061
Cash outflows							
Additional testing cost		300					
Future liability cost	25	25	25	25	25	25	25
Cooking facility investment			10,000	40,000	40,000	10,000	
Production line changes		3000	2000				
Original project cost (with risk)	20,000	30,000	40,000	3000			
Project progress risk			9				
Total cash outflows	20,025	33,325	52,034	43,025	40,025	10,025	25
Total net cash flows	18,862	32,162	45,871	30,862	19,862	20,087	30,036

costs may be incurred. On the other hand, if the project is ahead of its timetable, revenue can be achieved earlier and some cost savings may even be possible. Wisaforest has calculated the cost of the delay for each month delayed and the saving made from being ahead of the project timetable. By weighting these costs and savings with their respective probabilities, it is possible to calculate the expected net financial effect of the project progress risk.

The probability for the project:

- to stay on its timetable is 97.8%;
- to exceed the timetable by three months is 2.1%;
- to undercut its timetable by three months is 0.1%.

The assumed cost respective saving per each month:

- delayed is 158,000 mk;
- undercut is 395,000 mk.

The expected cash flow effect is therefore:

1999	2000	2001	2002	2003	2004	2005	2006	2007
35,000	40,000	40,000	40,000	40,000	40,000	40,000	40,000	40,000
35,000	40,000	40,000	40,000	40,000	40,000	40,000	40,000	40,000
25	25	25	25	25	25	25	25	25
25	25	25	25	25	25	25	25	25
34,975	39,975	39,975	39,975	39,975	39,975	39,975	39,975	39,975

$$(0.987 \times 0) - (0.021 \times 474{,}000) + (0.001 \times 1{,}185{,}000) = 8769 \text{ FIM}$$

The project was expected to be completed in 1994 and, therefore, the net effect was timed to that year.

EXPECTED MONETARY VALUE OF THE TOTAL INVESTMENT

Once each parameter had been identified, the respective uncertainty evaluated and the risk-weighted annual cash flows calculated, the EMV was calculated (see Table 4.3). The expected cash flows for each year were calculated as the difference between the cash inflows and the cash outflows. Then each net amount was discounted to year 0 using normal discounting techniques and the Kymmene corporate interest rate. This interest rate represents the weighted average cost of capital Kymmene Corporation has to pay for all its capital, both equity and borrowed capital.

The decision rule with EMV is that:

- if EMV > 0, go ahead with the investment;
- if EMV < 0 or = 0, abandon the investment.

COMPARISON BETWEEN ORIGINAL AND MODEL CALCULATIONS

Wisaforest calculated some financial indicators for the project based on its original expectations and figures (see Table 4.4). The calculations were not as detailed as they usually would have been in such a case, but gave some indication of the overall profitability of the project. The IRR was calculated to be 17.6% and the payback time without interest charges 6.9 years. On the basis of these figures, this investment was worth pursuing. The model calculation gave a positive EMV of 29.6 Mmk. The IRR was 14.9%. The EMV calculation would not have altered the actual decision.

TABLE 4.4

A comparison of the revenue and cost categories included in the original Wisaforest analysis and the model calculation

	Original analysis	New items
Revenues		
Operating margin	X	
Grant		X
Loan		X
Costs		
Testing	X (partially)	X (testing material)
Interest on TEKES loan		X
Storage cost, incremental		X
Market study	(in overhead)	X (N/A)
New common facilities		X
New waste management costs		X (N/A)
Investment expenditure	X (partially)	X (changes)
Future liability cost		X

The EMV calculation demonstrates the need to:

- define the project correctly — for each income dollar/pound/markka, there must be a relevant investment dollar/pound/markka spent;
- consider which parameters are the ones to which a probability needs to be assigned — only the large ones are necessary;
- clearly define the incremental items and their amounts to be considered — no general company interest charges should be included in the calculations;
- evaluate opportunity costs and gains.

The alternative calculation gives a different outcome due to the following main reasons:

- risk is incorporated in the calculation by assigning probabilities to some important variables;
- the effect of changes and expectations of changes in exchange rates are incorporated in the calculation;
- investments in changes required in the plant are incorporated in the calculation.

The main parameter that was not risk-weighted was the direct material cost of production. Timber prices are volatile and crucial to the profitability of pulp production. Volatility of pulp prices was understated although an exercise in assigning probabilities to price premiums was carried out. The underlying standard pulp prices were kept stable, which was an optimistic assumption for a period of 15 years.

The environmental aspects of the investment were dealt with at a differing level of consideration in the original calculations:

(1) The price premium to be gained from selling an environmentally-friendly product was under scrutiny. Wisaforest recognized that a price premium would disappear and a time-scale was attached to this in the original estimates.

(2) The regulatory environment both in the country of production and in the country of sale were studied both internally and with the help of external specialists. Authorities in Finland were contacted to check the present and future regulatory environment and its impact on permits and related costs.

(3) The safety of the process was studied using Hazop analysis and a risk factor was assigned to the new process. But the impact of this safety aspect was not taken into account in monetary terms in the original evaluations. The future liabilities were not assigned any monetary value and the impact on insurance premiums was not evaluated.

(4) Company image was a consideration in terms of progressing towards a better environmental performance. No evaluation was made of the financial impact

of changing a unit process and producing an environmentally-friendlier product. The financial impact was evaluated merely in terms of retaining a market and receiving a temporary price premium. The impact of company financing cost was not evaluated.

4.2.4   EVALUATION OF INVESTMENT OPTIONS

When making the decision whether to advance with the pilot plant investment, Wisaforest had many investment options; there was, however, no alternative to ozone bleaching technology for achieving the desired quality level. The choice originally made at Wisaforest and Kymmene Corporation built on some basic assumptions:

- the German market was important because of the price premiums obtained — even for standard pulp — and the TCF-based demand;
- TCF pulp would soon replace normal pulp in other markets as well;
- there was a price premium on TCF pulp that would compensate for the initial cost premium in TCF pulp production as opposed to STD pulp production. This price premium was expected to prevail until the year 2000;
- TCF pulp had to be produced by the Kymmene Corporation for strategic reasons, such as self-sufficiency and protection from supply constraint-based price peaks.

The perceived strategic nature of the investment in chlorine-free bleaching of pulp meant that calculations for other investment options were not made. In this chapter the investment options are discussed in so far as the other options could have been evaluated from an appraisal technique point of view. Wisaforest obviously considered other strategic options and related investments. The following sections make suggestions on which factors are crucial for the evaluation of these options.

PURCHASE TCF PULP FROM OUTSIDE SUPPLIERS

This option was abandoned as Kymmene Corporation sought to be self-sufficient in pulp production. Choosing this route would have meant strategic dependence on outside suppliers in market conditions where ability to produce TCF pulp was limited and hence provided a competitive edge to the producer. This could have meant occasional peaks in market prices and, potentially, undersupply of TCF pulp. Thereby the production of paper from TCF pulp would have been jeopardized and Kymmene's reputation as a supplier could have suffered.

The main aspects of evaluating this option would have been the estimate for demand-supply conditions in the market and the expected price level. Kymmene Corporation could have sold its standard pulp production through its normal channels, although some delay could have occurred in using these channels. In 1991 the Far East markets, such as Japan, were not attractive due to the strength of the Finnish currency. The cost of finding and tapping new markets would have been an important issue to consider.

WAIT FOR OTHERS TO DEVELOP THE TECHNOLOGY

If the TCF pulp market was considered sufficiently large, Wisaforest could have waited for others to develop a new technology and purchase it from them. In the meantime, Wisaforest could have purchased all TCF pulp needed from outside suppliers and sold its own standard pulp production to outsiders. The main considerations here are the price of a technology developed by others, the timing of such a purchase and Wisaforest's ability to adopt the new technology in a new plant. It is understood from discussions with Wisaforest that there might have been some negative cost effects of having to produce TCF pulp in an old standard pulp plant. This negative impact would have had to be estimated.

CHANGE TARGET MARKET

One of the basic assumptions in deciding to build the TCF pilot plant was the perceived necessity to remain geographically in the German-speaking market in Europe and to satisfy the demand for paper based on TCF pulp in this market. The German market represented some 55% of Wisaforest's pulp sales. There were two options to this market choice: looking for other markets and tapping a different market segment. The latter could have been achieved by producing pulp bleached using chlorine dioxide. The major consideration in this case would have been the price to be obtained for such pulp and the aggregate potential market demand.

BUILD A SMALLER PILOT PLANT

Wisaforest opted for a relatively large pilot plant that would have been able to meet the demand for ozone-bleaching for a quarter of the capacity of the hardwood pulp production. This made the investment financially heavy in the first three years and the annual cash flow remained negative. There could have been several reasons for this decision but one of them was technical: a large plant behaves differently from a laboratory or small plant. An option would have been to build a smaller plant that would have been less heavy financially and potentially would have yielded the same technical results. Had the experiment failed, less money would have been lost.

5. TEST CASE B — ICI ENVIRONMENTAL IMPROVEMENT PROJECT ON A UK CHEMICAL SITE

5.1 THE COMPANY

ICI is one of the major chemical companies in the world, with operations spanning the globe. The business activity involved in this environmental improvement project consisted of nine chemical production plants operating as part of a network on two sites. Their purpose is the manufacture of chlorine derivatives for ICI's Chlor-Chemicals business.

Output consists of chemical intermediates, plastics and solvents:
- chemical intermediates are used both on and off site as raw materials for other chemical processes;
- plastics includes plasticizers and PVC;
- non-flammable solvents are produced for use in a variety of end-market applications.

The existing plants have been operating for up to 30 years, and ICI regards this as a mature business activity.

5.2 HISTORY OF THE PROJECT UNDER STUDY

Manufacture of some of the chlorinated chemicals dates back to 1909 on the ICI sites which are involved in the study. Over the years the feedstock used to derive these chemicals has moved from coal-based to oil-based, as the underlying technology of production has changed.

The site is adjacent to a canal which runs into the river Mersey. The original method for disposal of aqueous effluent was by discharge to the canal, later incorporating the use of 'lagoons' to allow the settlement of solids, with evaporation of some components to atmosphere. The production plants currently in use were, by and large, designed and commissioned before environmental factors were as important as they are now. Whilst waste minimization was incorporated into their design, it did not meet today's criteria for environmentally acceptable chemical production processes.

In the early 1980s, a campaign to clean up the river Mersey was inaugurated, and ICI found itself subject to tighter 'consents' regarding its aqueous discharges from the site. This resulted in a continuous improvement programme being implemented on the site, focused on improving operating methods in the plants and increased use of recycling, rather than fundamentally changing the production methods used.

After the successful campaign by environmental groups, starting in the mid-1980s, against the discharge of chlorine and chlorinated substances, the European Commission issued new regulations controlling emissions to the aqueous environment. Within these were listed some 20 chlorinated hydrocarbons to be controlled or eliminated. Currently, detailed reviews and regulations have been issued on approximately 15 of the substances, and this impacts 75% of the production on these ICI sites.

The regulations can be applied in one of two ways:
- as Fixed Emission Standards (FES) — that is, with measurement criteria specified at the point of emission;
- as Environmental Quality Standards (EQS) — that is, with the measurement criteria specified for the 'receiving' environment — in this case the canal water standards.

Whilst most European countries have adopted the FES approach, the UK has opted for EQS. The effect for ICI is to require tighter standards of emission for its discharges to the canal than would be the case under FES. This is because there is very little natural dilution of discharges in the canal.

With the creation of Her Majesty's Inspectorate of Pollution (HMIP) and the passing of the Environmental Protection Act (EPA) in 1990, the situation under which ICI had been operating was radically altered. Regarding aqueous effluent, the previous regime of regulation by 'consents', which specified total annual discharge levels for pollutants, was augmented by Environmental Quality Standards, where the onus was on ICI to maintain the standard of the water to which the effluent was discharged.

For gaseous emissions, the situation was radically altered. Up to that stage, gaseous emissions had been a relatively unregulated area. It was brought into high profile with prescribed substances and a requirement to demonstrate that best available technology was being used to reduce or eliminate emissions.

Finally, under the EPA, HMIP was empowered to formally license the production units as fit to operate, in terms of their environmental impact. If HMIP rejected the improvement plans put forward by ICI for the treatment of waste on the site, ICI would literally lose its licence to operate. This set the scene for the investment review carried out by ICI in defining the scope, choice of technology and targets to achieve in its waste treatment project.

This case study looks at the situation when the project had been going on for 6–7 years. The incinerator was finally commissioned in the summer of 1995, almost ten years after the initial requirements were laid down by HMIP.

5.2.1   DEFINITION OF DECISION OPTIONS

ICI was faced with a situation in which:

● 'consent' levels for discharges from the site were being made ever more stringent, and applied more rigorously;

● there was an inability to meet internal ICI Group targets with the equipment and methods currently in use;

● an agreement reached with the NRA (who license the lagoon to be in operation) to close the lagoon by a specified date, had to be complied with.

There were three options:

● change the technology in the production plants to minimize the waste produced. Given the age of the units, this would amount to replacing the plants, which would mean an investment of hundreds of millions of pounds, and was not deemed economically viable;

● close the plants and terminate this business activity;

● treat the effluent produced to minimize the emission of regulated substances to the level of, or a level better than, the standards required under regulation.

ICI decided to implement the last option, though with concurrent waste minimization efforts in the production units where this was feasible.

The ICI Group policy was:

● to manage the company's activities to give benefit to society;

● to ensure that any new plants conform to the highest standards applicable in any country in which ICI operates that process;

● to reduce waste by 50% between 1990 and 1995.

It was this final policy directive which was the internal driving force for the investment under review.

ICI came to the conclusion that it would not be economically viable to build the plants anew. Was this the correct decision? Comparison of adding an end-of-pipe treatment to existing plant with investing in a totally new plant on an incremental basis gives an idea of the solution. In terms of the benefits — that is, cash flow in — the two investments should produce the same result, save for the new plant producing more efficiently and cost-effectively the same product. Hence it is the new investment required to get the product to the market that matters. Clearly, adding a waste treatment facility to the existing plant must be less costly than building a totally new plant.

In this instance, the waste treatment facility (or some similar investment) was required in order to obtain a continuing licence to operate from HMIP. Effectively a decision not to invest would have been a decision to leave this business area, since these production units were, in some cases, the only ICI plants making these products. The investment decision was not so much related to the achievement of a target return on net assets (RONA) from the investment, but was a case of answering the question 'Did ICI want to stay in this business area or not?'. If it decided to retain the business, the price for so doing was the cost of the investment required to ensure a continued operating licence from HMIP under the Integrated Pollution Control (IPC) regime.

In order to answer this question, a detailed review was carried out within the ICI project for each production unit and its related business to derive a view of the potential profitability and contribution to Group cash from each unit over the life of the proposed investment (standardized at ten years, though the design life is longer). The results of this exercise were incorporated in the investment case put to the ICI Board but are, of course, confidential to ICI. However, the overall conclusion was sufficiently robust to persuade the project team that detailed design and development work for the waste treatment facility should be undertaken.

It should be noted that, once agreement had been reached with HMIP on the standards which would apply and the time-scale in which they would be achieved, ICI was committed to making the full investment. Any performance not achieving the agreed standards would result in the licence being withdrawn. Once the investment was started it could not be rescinded unless there was a concomitant decision not to pursue this business. Nor would it be worth selling the production assets as a going concern, except to a buyer who wished to pursue the investment, or otherwise make the plant eligible for the licence.

5.2.2   CALCULATION FRAMEWORK

The standard measurement used at ICI is RONA over a ten year period. In this case, the production units were relatively old, and their book asset values were comparatively low. The expected life of the waste treatment plant was longer than ten years, but the investment case was standardized at a ten year time span. The businesses served by the production units were, by and large, in mature industries, where overall long term prospects were for a decline in market opportunities. The business area was classified as a cash generator, and investments were only justified on the basis of producing cash inflows in the short to medium term.

In the case of RONA, it is recommended that the replacement value of the assets is used rather than their present book value. Ten years can be used as the time horizon if it is clear that the annual inflows and outflows of cash will remain relatively stable after the first ten years. It is also correct if there are some items very far in the future, as their discounted value will be relatively minor. In this case, there was a strategic reason for choosing a ten year time horizon as the business was regarded as mature.

The assignment of site operating costs to the various business users required that the costs of the waste treatment investment be allocated to the different production plants. Costs needed to be allocated according to ownership of various activities. Whilst elements, such as direct piping from a production unit to the waste treatment plant, could be easily defined, there were issues over 'ownership' of joint outfalls and joint treatment processes to resolve.

This required not only an allocation of costs but the implementation of management processes to monitor and measure the effluent treatment and discharge, as well as to report and take action on the results of the monitoring. The operating paradigm was to treat effluent streams in the same way as chemical production process streams, and this has been one of the key 'attitudinal' changes brought about by the project to date.

The elapsed time of the project so far was between six and seven years in total, and this had brought the project team to the point of seeking local planning authority approval and ICI Board sanction. The total time expended to date was approximately 80 man-years. Elapsed time was roughly split into:
- two elapsed years to audit what effluents, at what levels of concentration, etc, were actually being produced on the site and discharged in various ways;
- two elapsed years to assess the different abatement technology options, to define the best technical solution;
- two elapsed years working up a detailed treatment facility design, together with the necessary changes in the production units.

In addition, a local planning authority application had to be prepared. This involved an assessment by ICI of the potential environmental impact of the new facility, which the planning authority then got reviewed by outside consultants. In addition, there had been public meetings to allow the local population to express its views. All this was time-consuming, both in terms of effort and elapsed time. This latter factor had been exacerbated by delays imposed by the council in pre-election periods, and the whole process had been ongoing for over two years. Simultaneously, detailed design was being done.

From sanction, it would be another 18 months or so before Phase 1 came on stream; in total, something like ten years elapsed time to implement a response to the initial requirements. The design has been modified over time to reflect changing demands, but it illustrates how much of an issue questions of future requirements are when the time-scales involved in implementing a response are of the order of six years.

The main decision aspects for the project were:
- what emission standards should be achieved;
- what technical solution would deliver these standards;
- what it would cost to achieve these standards;

● what would be the cost/benefit balance to ICI from directing its investment resources into this project rather than others.

5.2.3   PROJECT DEFINITION

After the strategic evaluation of the options for action, the actual project could be decided and defined carefully. All the strategic options were submitted to the same level of scrutiny so that the full implications of all alternatives were understood. The fact that some decision is deemed strategic does not mean that no calculations are to be made. Every strategic option has its financial profile and its risks.

The project, as defined by ICI, included the following elements:

● investment within existing production plants to minimize effluent production, pump aqueous effluents to a central storage tank and pipe vent gases together;

● new piping networks, including new pipebridges to transfer effluent to the central treatment plant;

● phase separation treatment of aqueous effluent using air stripping to remove chlorinated hydrocarbons (CHCs), followed by recovery of suspended solids by filtration;

● treatment of vent gases by thermal oxidation (incineration) to destroy CHCs and hydrocarbons (HCs). From this process there is recovery of energy as steam, and chlorine is recovered (as hydrogen chloride) before the treated vent gases are passed to atmosphere via a chimney.

This development, together with other site initiatives, would reduce site emissions of CHCs and HCs by 80% to the canal, and by over 90% to the atmosphere. In addition, the current method of treatment of aqueous effluent by passing it to a 'lagoon' would cease, and the lagoon would be closed.

Although this project focused on the technology issues of waste treatment and subsequent changes in the production plant, it is clear that there were many other issues that ICI needed to consider. These include the impact of the investment on company image and corporate environmental performance. The solutions to the technical questions require an R&D input in the form of technology evaluation. It was decided to leave the actual product and its design unchanged, and hence marketing issues were deemed not to be relevant. The product being a commodity, the additional cost had to be absorbed by the producer.

RESEARCH AND DEVELOPMENT

The R&D effort was focused on the technology issues of waste treatment. The technologies used, both in the production plants and in the waste treatment facility, had been chosen to ensure a high degree of waste minimization at source, together with reduced energy consumption. The waste treatment method provided the opportunity to recycle resources, partly by using the incinerator to provide energy for steam, and partly by recovering hydrochloric acid, which could be reused in the production process, thus reducing the need for production from other sources.

One of the key design criteria was to set the required level at which the technology would operate. The processes in these incineration units produce a level of dioxins in gaseous emissions. There is an EEC defined level for the emission of dioxins from incinerators (which defines the level at which there is assessed to be no significant threat to public health). There is also a common agreement within the industry on what may be regarded as 'Best Available Technology' for the treatment of such substances. Thus an initial analysis suggested that the ICI waste treatment plant should be based on achieving these levels, since they are defined as 'safe'.

It is expected, however, that through the lifetime of the investment technological advances will be made which may allow the achievement of even lower levels of dioxins in the emission plume. Also the regulatory authorities, reacting to pressure groups, may stipulate lower levels of emission in line with the ability of technology to deliver. However, there are three issues here:
- the ability to measure very small or zero levels of contamination;
- the moving target of what is a 'safe' level of emission;
- the necessity to prove that process and equipment designs actually work in practice.

The uncertainty facing the design team was whether to anticipate these potential moves in the design, or accept that further future investment would be needed if regulatory standards were revised. This was one of the key risk areas in the project. It was likely that future modifications would cost more to implement than if the features were designed into the waste treatment facility at the initial build stage. Modifications would involve shutdown of the waste plant and (potentially) production units whilst they are made.

The project team adopted the following rationale in addressing the issue:
- an assessment of the likely 'extra' cost making future modifications;
- a belief, based on past history of technically complex projects, that new working methods would be born out of the experience of working with the equipment, such that the original design performance parameters could be exceeded without significant changes to the existing kit;
- the possibility that advances in technology would make today's 'guess' at the future obsolete by the time it is to be implemented;
- an unwillingness to create precedents which regulators might follow.

94

The risk assessment process was discursive, rather than detailed. The cost of making all these assessments and necessary studies formed a part of the overall cost of the investment. Often companies do not include these costs in the investment calculation because there is no method for evaluating them in advance. They are often regarded as an operating overhead of a plant. Yet in a large project like this, where 80 man-years have been spent on preparing the non-detailed design aspects of the investment, the cost of those years is a relevant and significant part of the total investment cost.

PRODUCTION PROCESS

The total cost of the waste treatment phase of the project was some tens of millions of pounds. The major item required in the waste treatment process was the incinerator unit. Whilst this was one third of the total cost of capital equipment required, in fact it represented only 20% of the total project cost. Approximately 50% of the project cost was not the cost of the waste treatment plant itself, but the cost to 'hook-up' the production units to the plant, so that their waste could be collected in a central location for treatment.

This is often the case in 'end-of-pipe' investment proposals. The cost of the changes required to existing buildings, with new piping and ducting, and perhaps the need to move equipment, is a significant proportion of the total cost. The cost of treatment facilities cannot be estimated on the basis of the capital cost of the treatment equipment alone.

The waste treatment plant would need monitoring and maintenance. These costs are often surprisingly high and are usually omitted from operating cost estimates. Internal financial monitoring does not provide easy access to these costs in many companies. Some companies have therefore started campaigns to alter their accounting procedures so that site managers have a better overview of the environmental costs of the plant.

COMPANY IMAGE

Within this project local concerns were addressed and specific statements made in the planning documents on:
- visual impact;
- effect on flora and fauna inhabiting the site;
- local water and air quality;
- local geology and hydro-geology;
- effect on road traffic outside the site once the project was constructed;
- any change in noise levels.

Also a Public Health Impact Study was made, though this is not (yet) a formal requirement for planning consent.

Obviously, the detailed review of each of these factors was in order that a statement could be made on the likely increases in cost of the planning and design stage of the project. However, such work was necessary if ICI was to convince the local population, and the planning committee of the local council, that the investment should be allowed to proceed. The cost of performing this work was an integral part of the overall investment cost.

A further substantial investment in company image was made in the project for equipment to be installed to prevent visible condensation of water vapour in the plume from the incinerator stack. Without this equipment, white vapour would be seen issuing from the stack. This vapour would in fact be steam. Because of public sensitivity, the project team decided to install equipment which would stop the moisture issuing as steam. This equipment has no effect on the level of toxic substances venting from the stack. Subsequently, ICI was informed by HMIP that the equipment was required before consent would be given.

The capital cost of this equipment was approximately 2% of the total capital budget, but the operating cost is about one third of total operating costs. This is a significant cost burden for an item which is included essentially only for its cosmetic effect. No actual estimate was made of the financial return of the good public image gained from this investment.

5.3   POTENTIAL FURTHER USE OF THE METHODOLOGY

5.3.1   EMISSION STANDARDS

The particular aspect of dioxin emissions was a central issue and the installation of a steam plume suppression fan in the vent stack to prevent the appearance of white vapour from the stack involved large costs. No quantitative risk analysis was applied to either of these decisions. Using the methodology, each of them could have been broken down into component factors, with each factor being assessed and a risk factor assigned.

For example, knowledge of the legislation currently in force, and its possible future direction, could be applied to review what future control standards are 'needed'. The risk-weighted cost to build these into the current design could be calculated — that is, the EMV of the possible outcomes. This EMV is the basis for a comparison of the cost of building potential future requirements into the current design, and the cost of having to build new facilities or refit items in the future.

96

The calculation could be done for each potential level of regulation, to see the likely future cost. The result could be shown in a matrix (see Table 5.1).

The EMV can be compared with a similar set of calculations which review the cost of making a change to the waste treatment unit in the future at each potential level of change. Then a comparison will show what the 'trade-off' is between speculating now, and waiting for the future to become more certain.

For the investment in the steam vapour suppression equipment, the methodology would highlight the need to compare the costs of this equipment with quantified analysis of potential losses (from impaired public image) if the equipment were not installed. It might reveal that money spent on educating local residents about the steam vapour plume would be more cost-effective than the capital and operating costs of the equipment installed to suppress the plume. Once the steam vapour suppression equipment became a demand from HMIP, the return on this investment would be the net present value of future sales. The equipment effectively secured the licence to operate.

TABLE 5.1
Calculation of likely future costs (EMV)

Potential level of regulation against today's standards	Cost to achieve required level of emission, if designed into plant now	Probability of occurrence/possibilities of each outcome	Expected monetary value (cost) to implement change to design of waste plant
Small reduction in emissions	£A1@ £A2@ £A3@	most likely 10% chance to exceed 90% chance to achieve	£A
Medium reduction in emissions	£B1@ £B2@ £B3@	most likely 10% chance to exceed 90% chance to achieve	£B
Large reduction in emissions	£C1@ £C2@ £C3@	most likely 10% chance to exceed 90% chance to achieve	£C

5.3.2 SELECTION OF TECHNICAL SOLUTION AND THE COSTS OF IMPLEMENTATION

The methodology could be used to highlight factors of difference between technical options in aspects which may not usually be considered — for example, the effect of a particular choice on the insurance risk rating for the operation. Whilst not impacting on the purely technological assessment of which is the most 'elegant' technical solution, in all cases the 'best' solution is based on a mix between its technical excellence and the economics of implementing it. It is in the economic assessment that the methodology can improve current methods. This is through the inclusion of a broader range of factors with assessment of risk for key items of uncertainty. Some of these factors have been mentioned in the relevant parts of the project description.

In its consideration of the various technical options, ICI could have applied the principles of the methodology to each option to gain a fuller understanding of the key risk drivers in the selection criteria. This would identify which ones needed further investigation to reduce their uncertainty, and thus that of the project as a whole. It is possible that, in the desire to guarantee achievement of current regulatory standards, some aspects of the plant are 'over-engineered' and risk assessment would have initiated lower cost solutions. The methodology would also identify cost/benefit balances in the choices made on, for example, using a waste water total containment system, rather than allowing surface water to run off the site.

5.3.3 DIFFERENTIAL RETURNS FROM POSSIBLE INVESTMENTS

As with all companies, ICI has finite investment capital available to allocate to potential investment projects generated by the component businesses. In the test case the investment is not, on the face of it, specifically revenue-generating. The costs of the waste treatment facility will add to the cost base of the businesses involved, without directly generating offsetting revenue of the same order, although some benefits are gained from the steam generation and hydrogen chloride recovery.

If the analysis is extended to look at the alternative scenarios, a different picture emerges. ICI has already done this to some extent in its review of the future prospects for the production units. The company reached the conclusion that these operations are worth preserving, even with their extra cost base, rather than closing them down when their operating licence is refused. The methodology suggests areas for investigation where the full benefits of the investment can be made apparent, especially in trying to put a value on the improved public image of the site.

5.4   SOME CONCLUSIONS FROM THE REVIEW

The test case highlights the necessity of evaluating financially all strategic options available to a company. Cursory evaluations are not adequate even when a decision is perceived as merely strategic — quite the opposite. The model and the methodology assist companies in breaking down the overall strategic situation. This enables companies to assign risk factors to each component instead of having to evaluate the overall risk of the whole project.

The need to respond to public demands often requires expenditure on company image as well as on technology — for example, the plume suppression equipment. Yet it is necessary to gain an understanding of the financial cost of such an investment and the actual return of it. Public opinion may not be relevant in the short term from a licence-to-operate point of view but in the long term it will affect the public perception of the industry overall.

Every situation is site-specific — local (and state and national) regulations differ (and sometimes conflict). Every plant is different and the cost of end-of-pipe technology differs greatly from site to site. The choice of a technology — be it a waste treatment plant or an intrinsically cleaner production process — depends on the local parameters. The cost of fitting new equipment into existing operations can double the costs of the project, because of the need to carry out civil engineering projects for 'tie-ins' to provide the infrastructure for waste treatment.

In general, the actual cost of licences is trivial compared with the cost of monitoring which is required to gain and maintain licences. Companies in Britain are usually experiencing problems with the collection of relevant ambient and emission data when preparing their IPC applications. This often means that less resources are available for actual improvement of the environmental performance of a site.

The elapsed time in moving from acknowledging the requirement for change to actually implementing new methods or equipment to effect that change can be significant. Also a large amount of internal effort is likely to be necessary to assure the regulatory authorities that sufficient progress is being made in improving standards. These costs are often significant and, unfortunately, often omitted from investment calculations. This case clearly highlights the importance of these costs.

APPENDIX 1 — USER GUIDE TO THE METHODOLOGY

A1.1 INTRODUCTION TO THE CHECK-LISTS

The check-lists are divided into the following modules:

- research and development;
- product design;
- production;
- marketing;
- company image.

This structure is illustrated in Figure A1.1.

For each module the check-list covers the following items:

- definition of the module boundaries;
- relevant out-of-pocket costs;
- relevant revenues;
- risks of each parameter and associated probabilities;
- sources of data on each parameter.

The modules represent the areas of responsibility through which a product progresses. In each module, a product brings costs, returns and risks, and they influence the financial outcome of the company. The check-lists in

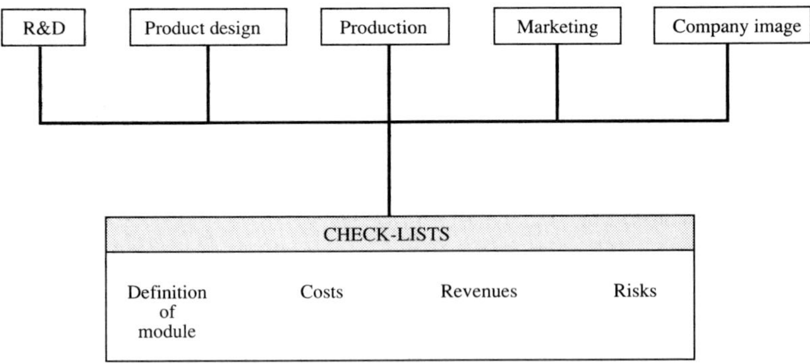

Figure A1.1 Structure of check-lists for each model.

100

each module help analyse these costs, returns and risks. The following definitions apply:

COSTS

Costs are the investments, expenditures, expenses or out-of-pocket cash flows that amass when a company undertakes the investment in question. The costs are defined as future cash outflows. Non-cash items, such as depreciation, are not included. Also sunk costs, such as past R&D costs, are not included.

RETURNS

Returns are financial benefits, revenues, profits or cash inflows that accrue to a company undertaking the investment in question. The benefits are defined as future cash inflows.

RISKS

Risks are defined as the uncertainties attributable to the possible values of various cost and benefit items included in the analysis. Risk is incorporated in the form of probabilities and not in the form of contingencies or discount rates that contain a 'risk premium'. The outcome of each possible value of a parameter is defined by a probability distribution.

To assist in the technical calculation of the project outcome, refer to the test case examples in Chapters 4 and 5. Appendix 2 explains in detail how expected monetary value (EMV) is calculated. Environmental aspects are marked with an asterisk. Note that some of the items may have environmental aspects although these are not marked separately.

Use either PC-based programs — such as Lotus, Excel, or various risk management programs — or a calculator and a piece of paper to calculate the outcome. If many parameters are included, a computer-based program may be necessary. Probability distributions for a parameter are best calculated outside the model, as the model may otherwise become very slow to run through.

LCA, Hazop/Hazan and other analyses providing insight into the risk and monetary value of relevant parameters are assumed to have been carried out outside the model. Only the necessary information from these analyses will be incorporated into this calculation.

A1.1.1 SOME POINTS TO WATCH

DISCOUNTING MAY NOT REFLECT PROJECT RISK

The most common mistake in applying discounted cash flow techniques is the incorrect choice of discounting factor. The reason may lie in the original definition of the discounting factor. The factor should reflect the risk that capital markets assign to a company. It therefore reflects the sum of the risks a company takes, and includes various projects with their respective risks.

Each factor in an investment appraisal has a different risk and uncertainty. Assigning one risk factor, the discount rate, to all of them means that some risks are overstated, others understated. If a wrong risk factor is assigned to a significant cash flow, such as sales, the whole calculation can be distorted.

UNCERTAINTY CAN BE DEALT WITH!

A factor is often omitted because quantifying and estimating its value is deemed impossible or very difficult. In most cases, three steps in structuring uncertainty can help to solve the problem:

(1) List all the factors that bring uncertainty to the value of a parameter.

(2) Choose which factors are important.

(3) Estimate the probability distribution, based either on experience or simulations, of the potential values for each factor to be reviewed.

The process may seem oversimplified, but it works in practice. Specialists in various areas of technology or markets gain a good 'feel' for the behaviour of those areas. This 'feel' can mostly be interpreted as a probability distribution as, for example, in the Finnish pulp mill (see Chapter 4).

USE OF LIFE CYCLE ANALYSIS HELPS

Despite many flaws, one of the major advantages of LCA as a methodology is that it can help a company to list the various types of impact a product or a process has on the environment. It may be difficult to estimate the monetary value of this impact but, if the company knows the impact in size and type, it can try to estimate the future potential likelihood of an event. The cost of such an event can then be estimated either directly, based on experience, or indirectly by using a proxy. For example, potential future liability costs may be of such a size that they make an investment unattractive.

OVERHEADS HIDE MANY THINGS

Company cost accounting is usually based on directly attributed costs and overheads. Overheads usually include:

- marketing costs;
- purchasing costs;
- administrative costs;
- depreciation.

The group of administrative costs is the most general one and hides many costs items that actually can be attributed to a department or project. Only the tradition of including these costs in overheads prevents companies allocating them correctly. Many of these administration costs can be attributed to a project in the appraisal phase.

THERE IS NO SUCH THING AS A STRATEGIC DECISION

When starting to evaluate an investment a company may be faced with various alternatives. The temptation is to regard some of these as less strategic than others and therefore not to analyse them. There is a warning here: just because a decision is strategic does not mean that the full implications are unimportant. A strategy based on product or market demand may save the company in its markets but the financial implications of the strategy may be disastrous. The financial implications may also depend on such unpredictable elements as commodity prices or exchange rates. This unpredictability needs to be incorporated in the financial analysis of any product- or market-based strategy.

DO NOT JUMP TO CONCLUSIONS

When faced with a new investment and the evaluation of its implications, the company should base the decision on its overall product, environmental and business strategy. These strategies may be based on some 'old' truths and assumptions about cash generation ability or market share. As the investment under consideration may reach far into the future, the underlying strategic assumptions behind it and the alternatives to it must be re-evaluated.

A1.1.2   STEPS IN USING THE METHODOLOGY

STEP 1: DEFINE THE DECISION OPTIONS

There may be several options for the use of investment funds. Define what they are and what information about these options is needed. The options are really choices between different ways of achieving a company goal. The options can therefore be alternative technologies or completely independent investment alternatives.

STEP 2: DEFINE THE PROJECT UNDER STUDY

The project may be an R&D project or an investment proposal for building a new plant or amending a production process. It is important to identify the project focus in order to select the right information sources and issues to review.

STEP 3: COLLECT BACKGROUND INFORMATION ON THE PROJECT

Each project needs to fit into the company strategy. Find out what the company strategy is to understand why this project is important for the company. List the issues for and against this project in terms of company strategy and culture. Be aware of what is expected from an investment appraisal and what is required if an investment is to be accepted.

STEP 4: IDENTIFY THE PEOPLE INVOLVED AND INFORMATION SOURCES REQUIRED

A project may be proposed by one department but will affect the work and decisions of other departments or other parts of the company. Understand who will be involved in and affected by this project. They are the primary sources of information. Consult outside specialists early on in the project appraisal process.

STEP 5: CARRY OUT THE ACTUAL APPRAISAL WITHIN THE FRAMEWORK ESTABLISHED IN STEPS 1 TO 4

(a)   Decide whether to make the calculation on an incremental basis or not.

(b)   Choose the time horizon for the appraisal.

(c)   Using the check-lists, choose only those areas that have a future cash flow effect on the company. Disregard sunk costs (that is, expenditure in the past). Analyse cost items that are usually regarded as overheads and direct them to your project. Beware of double-counting. To gain more in-depth understanding of the check-list items, consult Chapter 2.

(d)   To find relevant data on each parameter contact relevant parts of the company and outside professionals and specialists as appropriate.

(e)   Estimate probabilities for each possible outcome of each parameter to be considered — either alone or with the help of specialists.

(f)   Build the NPV and EMV equations. The equations can be run either on a calculator, on risk management software, or on a PC application such as Lotus or Excel.

(g)   Calculate the outcome, either as one NPV or several under different scenarios.

(h)   Calculate the EMV of the investment.

The elements of Step 5 are summarized in Figure A1.2.

A1.2   GENERAL ASPECTS OF THE APPRAISAL

(1)   The calculation basis can be one of two methods — incremental or separate projects compared.

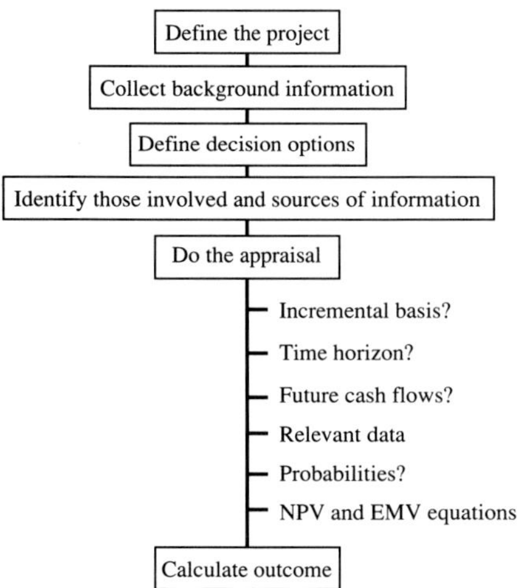

Figure A1.2 Steps in using the methodology.

Incremental calculations are appropriate when the project involves a relatively minor change in present circumstances — for example, a new unit process or a new product feature. Comparing independent projects is relevant when the project under study is totally new to the company — that is, not changing parts of existing circumstances, or where there are several totally independent projects competing for the same funds.

Incremental   ☐
Separate projects   ☐

(2*) Fixing the time horizon for project appraisal.
- If a company rule exists, use that.
- If company recommendations exist, evaluate those.
- If the choice is free, use the most appropriate time horizon.

Many companies set rules for investment appraisals and techniques to be applied. Often various types of investments are to be evaluated in a set time horizon. If no such rules exist, there is a free choice. The time to be considered depends on the physical life of potential assets or the expected lifetime of the product to be launched. Often the final years of a long time horizon lose relevance due to discounting. In these cases the important cash flows can be converted into perpetuities.

Time horizon   . . . . years

(3) Choosing the relevant discount rate.

• If a company rule exists, use that.

• If a company recommendation exists, evaluate that.

• If no recommendation exists, consult finance or treasury departments for information.

Remember that the discount rate is to reflect the opportunity cost of capital, ie the best alternative return available for the money that is to be used for this project. The discount rate shall *not* reflect project risk although it does reflect company risk. The latter is determined by the capital markets and is the cost of financing to your company. Project risk is incorporated by using the probabilities of each possible outcome of each parameter considered in the equation.

Discount rate   . . . . %

(4) Deciding how to treat overheads. It is easier, but less accurate, to use company overhead rates rather than analysing individual costs. Remember that these can hide large cost items that are directly attributable to the project and should therefore be taken into account. The actual cost accounting of the company does not have to follow the same rules for overhead allocation as the project appraisal. Beware of double-counting or forgetting some items when there is deviation from company overhead definitions.

Items cut out from normal overhead definitions:

Item	*Amount*
. . . .	. . . .
. . . .	. . . .
. . . .	. . . .

New overhead percentage for other overheads   . . . . . .

(5) Many companies still use the division of costs into fixed and variable. These are relevant only in very stable circumstances. The project may change only one unit process but it may have major repercussions on other parts of the company (see Test Case A on page 63). Forget about the division into fixed and variable costs and take into account all costs that depend on the project.

(6) Each parameter affecting the project must be analysed for the possible outcomes and the respective probabilities of these outcomes. A normal distribution of probabilities can be assumed initially if there is no reason to think that the distribution is different. Look at the discussion in Appendix 2.

106

CHECK-LIST A — RESEARCH AND DEVELOPMENT

IDENTIFICATION OF MODULE BOUNDARIES

1 How is the R&D project under study associated with the project for which the overall appraisal is being made?

(a) This is the project itself ☐

(b) This is specific R&D done for the project ☐

(c) This is general R&D done in the company and
only loosely associated with the project ☐

In (a) and (b) the project is clearly identifiable. In (c) the cost and revenues associated with the overall project being studied need careful identification and separation from the general R&D costs and revenues. Beware of including 'sunk' costs — that is, money already spent in the past. Only future cash flows are relevant for the appraisal.

Notes:

2 What is the goal of the R&D project under study?

(a) Cost reduction, either product or process ☐

(b) Profit sustenance, either product or company ☐

(c) Quality improvement of product ☐

(d) New product development ☐

(e) New process development ☐

(f) Development of saleable technology ☐

(g) Pollution prevention/reduction ☐

Although the overall R&D project may have several goals, break it down to components so that each goal has its own costs and revenues. If the overall goal is non-environmental in the first instance, separate out any environmental aspects. This allows the company to find out how non-environmental and environmental projects compare in cost and revenue terms. The comparison is needed when both types of investment compete for the same total investment funds available.

Notes:

3* If the R&D project is environmentally focused, the ultimate goal could be:

(a) Emission reduction ☐
(b) Reduction in energy use ☐
(c) Reduction in raw material use, one or several ☐
(d) Development of process internal recycling ☐
(e) Development of recyclable end-products ☐
(f) Reduction of health risks from product ☐
(g) Reduction of health risks from production ☐
(h) Extension of product durability ☐
(i) Improvement of environmental performance
of company ☐
(j) Lowering risk for liability costs ☐
(k) Lowering overall operational risks ☐

When the end goal of the R&D project is identified, it provides a good guide for identifying revenue streams and potential sources of risk. Often the environmental goal serves many purposes simultaneously and in such a case there may be one primary goal and others that follow from this goal.

Notes:

4 **What is the source of R&D associated with the overall project being appraised? For example:**

(a) Outside contractors ☐

(b) In-house R&D ☐

(c) Both ☐

Even though the actual technical R&D may be done by outside contractors, there are many tasks that can still be performed by in-house staff and linked directly to the project under study. The in-house staff may be involved in drafting specifications or managing the outside contractors, for example. The company may also be involved in financing an outside research project that contributes to the project under study. Identify these costs because they may be significant. The division of R&D work between in-house staff and outside contractors is relevant when trying to identify the costs for R&D.

Notes:

5 What is the role/function of the R&D department in the company? For example:

(a) To carry out R&D only ☐

(b) To monitor environmental legislation ☐

(c) To make recommendations for the environmental policy of the company ☐

(d) To formulate company environmental policy ☐

(e) To make strategic product decisions without marketing department ☐

(f) Other (specify)

. . . .

. . . .

. . . .

The purpose of the R&D department can vary from company to company. Depending on the role, certain costs and revenues are associated with this department or other departments — for instance, legal monitoring and fact-finding costs. This question and the answers to it provide guidance when trying to break down the R&D departmental overhead and when incorporating all relevant overheads from other departments. Many companies either ignore these costs or underestimate the total amount of time spent on these issues.

Notes:

6 If this is the actual investment to be evaluated (ie, R&D is not just a part of another investment), make a list of the persons involved and how other parts of the company — other departments, subsidiaries or sites — are affected by this project. The project may have direct cost and revenue consequences for these organizational entities and these may need to be attributed to your investment. For instance, a sales company abroad may charge your unit for sales work done on your behalf. Include this cost in the investment appraisal.

Notes:

COSTS

COSTS RELATED TO R&D

1 Direct labour cost

(a) Researchers   . . . .

(b) Technicians   . . . .

(c) Testing staff   . . . .

(d) Laboratory staff   . . . .

(e) Other directly involved staff   . . . .

Estimate man-hours and calculate total cost for this labour. The role of managing the project can be significant and therefore can add to the direct costs of the project. Although in many companies management is included in overheads, for the purposes of the project cost appraisal it is relevant to separate out the management cost. This is particularly true if environmental legislation is to be monitored by R&D management. Do not underestimate the time spent on this activity.

Notes:

2   Indirect labour

(a) General overhead labour (if overhead rate is defined as such)
Overhead percentage   . . . .
(b) Specific accounting, marketing and other departments' labour who spend time specifically working on this project (watch out for double-counting)

Department	*Costs*
. . . .	. . . .
. . . .	. . . .
. . . .	. . . .
. . . .	. . . .
. . . .	. . . .
. . . .	. . . .

This is usually the most difficult area to estimate because the cost accounting method used by many companies seldom supports this analysis adequately. Separating out these costs may require assistance from accounting and other departments. Other areas of cost analysis are given in sections 3 to 7.

3	**Direct equipment required**	. . . .
4	**Overhead not yet included**	. . . .
5	**Cost of outsourcing**	. . . .
6	**Cost of testing and associated waste management**	. . . .
7	**Cost of patents and licences (product or technology)**	. . . .

Notes:

COSTS IN OTHER PARTS OF THE COMPANY DUE TO THE RESULTS OF R&D (watch out for double-counting)

1 Cost of investing in new assets required
(see Check-list C) . . . .

**2 Operating costs of new investment required due
to results of this R&D project**
(see Check-list C) . . . .

**3 Estimated cost of new marketing required in all parts
of the company, also overseas, due to this R&D project**
(see Check-list D) . . . .

4* Estimated future liability costs
(see Check-list C) . . . .

5* Estimated monitoring costs
(see Check-list C) . . . .

6 Costs of patents and licences required . . . .

**7 Cost of procurement for changes required
in other parts of the company** . . . .

**8 Cost of financial risk coverage — for example,
options or futures** (see Check-list C) . . . .

**9 Cost of tasks performed in other parts of the
company due to this project** . . . .

COSTS IN INCREMENTAL CALCULATION

If the calculation is to be made on an incremental basis, calculate the costs according to categories 1 to 9 but only on an incremental basis. This means that only costs that occur due to this project are taken into account. It is possible that there is an alternative that achieves the same goal as this investment proposal. Include the costs occurring due to the alternative in the calculation.

REVENUES

REVENUES FROM THE RESULTS OF R&D

1 Estimated sales volume due to this R&D project
(see Check-list D) . . . .

2 Revenue from sales of patents or licences . . . .

3 Revenue from grants and subsidized credits . . . .

4 Revenue from prizes and awards . . . .

**5* Revenue from sales of recycled material
and components** . . . .

6* Revenue from recycled by-products . . . .

REVENUES IN INCREMENTAL CALCULATION

**1 Estimated savings in operating costs due to this
R&D project** (see Check-list C) . . . .

**2* Estimated reduction in future liability costs due
to this R&D project** (see Check-list C) . . . .

**3* Estimated increase in revenues from sales of
recyclable materials** (see Check-list C) . . . .

**4 Estimated savings on reduced insurance costs
and financing costs** (see Check-list E) . . . .

**5* Estimated savings on achieved regulatory
compliance** (see Check-list C) . . . .

**6 Estimated increase in sales due to this
R&D project** . . . .

Notes:

RISKS, ASSOCIATED PROBABILITIES AND DATA SOURCES

1 Technical risks

(a) Estimated performance level not achieved
(b) Inadequate performance level achieved
(c) Complete failure of project

Each of these potential outcomes of the project needs a probability assigned to it. To derive this, each risk category can be broken down into details. These risks can be estimated by the R&D staff. The level of risk depends greatly on how large a change is being planned for.

Description of the particular risk: . . . .

. . . .

. . . .

. . . .

Description of measure and source of data: . . . .

. . . .

. . . .

. . . .

Expected monetary value (EMV) weight: . . . .

. . . .

Notes:

2   Progress-related risks

(a) Inadequate manning
(b) Inadequate time allowance
(c) Unforeseen obstacles
(d) Inadequate project management
(e) Inadequate funding

The probability of these occurring depends greatly on the project organization. Each company has different levels of skills in project management and different levels of experience in R&D work. It is up to the R&D and company management to evaluate the level of uncertainty in this respect. Again, break these risk categories down and give further details, if appropriate.

Description of the particular risk: . . . .
. . . .
. . . .
. . . .
Description of measure and source of data: . . . .
. . . .
. . . .
. . . .
Expected monetary value (EMV) weight: . . . .
. . . .

Notes:

3*   Risks related to moving targets

(a) The regulatory limits change
(b) Company policy changes
(c) Market expectations change
(d) Other internal expectations change
(e) Other external expectations change

Regulatory limits are bound to change but the speed at which they change
needs to be estimated. It can also be company policy to perform better than
the regulatory limits. Prediction of changes in regulatory limits can be done
in a number of ways. The general prediction may be by reviewing:
• if the substance under consideration is already strictly regulated in most
countries, the change can be rapid;
• if the industry is closely lobbying the regulatory bodies, the change can
happen more slowly and the lobbying group can give valuable information
on this;
• if the substance is not yet being regulated, the change can be sudden.
Pressure groups can lead the way in bringing new substances under scrutiny.

The judgement for predicting changes in regulatory limits is subjective
and should be made by specialists who monitor the regulatory environment
closely.

Company internal policies may change. These predictions are subjective
and should be made by the R&D management and technical management
together. The prediction is based on the company's ability to move in a new
direction — for instance, to a stricter environmental policy. This ability
depends on the financial restrictions of the company.

In Test Case A (see page 63) the company policy was to be self-sufficient in
pulp. Therefore the non-chlorine bleached pulp had to be produced by the
company and not purchased from outside. Investment in this technology was
required and, had the company changed its internal policy, purchasing from
outside could have become an option.

Market expectations may change radically due to accidents or other
incidents, or successful lobbying by various interest groups. These
expectations need to be monitored by the marketing department
(see Check-list D).

Risk source	*EMV weight*
. . . .	. . . .
. . . .	. . . .
. . . .	. . . .
. . . .	. . . .
. . . .	. . . .
. . . .	. . . .
. . . .	. . . .
. . . .	. . . .
. . . .	. . . .

Notes:

CHECK-LIST B — PRODUCT DESIGN

IDENTIFICATION OF MODULE BOUNDARIES

1   What is the nature of the project that is being evaluated?

Project is to:
(a) Alter a product design   ☐
(b) Create new production process   ☐
(c) Develop a new product   ☐
(d) Improve company sales   ☐
(e) Other (specify)

. . . .

. . . .

. . . .

This helps identify the specific parts of the company that are to be involved in the project. These areas will provide the necessary information and the project must take into account the cost and revenue implications of product design in all these areas.

Notes:

2 What is the goal of the design work?

(a) Improve the quality of an existing product   ❏

(b) Improve marketability of an existing product   ❏

(c) Improve environmental aspects of a product   ❏

(d) Improve the production process of an
existing product   ❏

(e) Other (specify)

. . . .

. . . .

. . . .

Identifying the goal of the design work helps to focus the emphasis in cost and revenue accumulation. If, for instance, it is marketability that is to be increased, there must be an idea of the potential selling price and new volume, and hence the budget for design work can be put in place. If it is the production process that is problematic because of present product characteristics, there must be an estimate of what the savings could be if the process were changed and hence what the design budget could be in total.

Notes:

3* If the goal is to improve the environmental performance of a product, what is to be achieved by the product design changes?

(a) Ruggedness of product ☐
(b) Multi-purpose product ☐
(c) Reliability of product ☐
(d) Non-toxic/less toxic product ☐
(e) Ease of manufacture ☐
(f) Lighter weight product ☐
(g) Ease of servicing ☐
(h) Standardized composition ☐
(i) Recyclability increased ☐
(j) Other (specify)

. . . .
. . . .
. . . .

Prioritize what is to be achieved — that will direct the activities of other departments and hence the costs accruing in these departments if a new design is being created or an old one changed. For instance, if the design is to be changed because of waste treatment problems, changes in the company caused by this decision can be:

• in production, where new assets need to be purchased, safety measures changed, staff trained and so on;

• in procurement, where new procurement contracts need to be put in place, and new currency and commodity markets observed and monitored;

• in accounting, where new cost accounting may have to be designed and accounting programs altered;

• in marketing, where new marketing campaigns may need to be planned, and new markets studied;

• in public relations, where environmental information packages need to be updated.

All these changes accumulate costs to the company. Although the present cost accounting system and overhead definitions do not support the identification and separation of these cost items, it is essential for the company to try to estimate these costs in the project. They are often significant and frequently hidden.

COSTS

COSTS DIRECTLY RELATED TO PRODUCT DESIGN

1 Direct labour

(a) Designers . . . .

(b) Assistants to designers . . . .

(c) Testing staff . . . .

(d) Project management
(separate out from overheads) . . . .

(e) Other staff directly involved . . . .

Estimate man-hours and calculate total cost of these staff. Project management can be costly because the project manager may have to liaise extensively with other parts of the company and outside parties. Although present cost accounting and overhead definitions may not support this separation, it is important to do it when performing the appraisal, if the amounts are significant.

Notes:

---

2 Indirect labour

(a) General overhead (if not included elsewhere)   . . . .
(b) Specific staff involved in product design in other
      departments if not already included
      (preferably separate these costs from general
      or departmental overheads)   . . . .

---

3 Direct equipment required

Item	*Costs*
. . . .	. . . .
. . . .	. . . .
. . . .	. . . .
. . . .	. . . .
. . . .	. . . .
. . . .	. . . .

---

4 Overhead not yet included   . . . .

---

5 Cost of outsourcing   . . . .

**6* Cost of testing and associated
       waste management**   . . . .

**7 Cost of patents and licences
     required for product design**   . . . .

Notes:

COSTS IN OTHER PARTS OF THE COMPANY DUE TO PRODUCT DESIGN (watch out for double-counting)

1 Cost of investing in new assets which are required due to design change
(watch out for double-counting)   . . . .

2 Operating costs of new production investment required due to the design change
(see Check-list C)

(a) Raw materials (change in cost or amount, if comparing with an old product)   . . . .

(b) Energy use (change in energy use, if comparing with an old product)   . . . .

(c) Quality performance (change in number of defects/rejections)   . . . .

(d) Waste treatment (change in treatment required)   . . . .

3 Estimated cost of marketing
(see Check-list D)   . . . .

4* Estimated future liability costs
(see Check-list C)   . . . .

5* Estimated monitoring costs
(see Check-list C)   . . . .

6* Estimated cost of services to customers
(a) Packaging   . . . .
(b) Field service network   . . . .
(c) Spare parts   . . . .
(d) Recycling   . . . .

7 Cost of procurement required due to new design   . . . .

8 Cost of financial risk coverage — for example, options or futures
(see Check-list C)   . . . .

COSTS IN INCREMENTAL CALCULATION

1 Estimated reductions in present production levels and sales volumes due to design changes (see Check-list C)

Volume   *Sales*

. . . .

. . . .

. . . .

. . . .

. . . .

. . . .

2 All other cost types mentioned above, but on an incremental basis

Cost   *Incremental change*

. . . .

. . . .

. . . .

. . . .

. . . .

. . . .

Notes:

REVENUES

REVENUES FROM PRODUCT DESIGN

1 **Estimated value of new sales**
(see Check-list D) . . . .

2 **Revenue from sales of patents or licences** . . . .

3 **Revenue from grants and subsidized credits** . . . .

4 **Revenue from prizes and awards** . . . .

5* **Revenue from sales of recycled material
and components** . . . .

REVENUES IN INCREMENTAL CALCULATION

1 **Estimated value of increase in sales** . . . .

2 **Estimated savings in operating costs
in production** (see Check-list C) . . . .

3* **Estimated savings in future liability costs**
(see Check-list C) . . . .

4 **Estimated savings in insurance and
financing costs** (see Check-list E) . . . .

5* **Estimated savings in achieved regulatory
compliance** (see Check-list C) . . . .

6* **Estimated increase in sales of recyclable
materials and components** . . . .

Notes:

RISKS, ASSOCIATED PROBABILITIES AND DATA SOURCES

1 Changes in the regulatory environment

(a) For the end-product and its
     components (see Check-list D)   ❑
(b) For the production process
     (see Check-list C)   ❑
(c) For key raw materials (see Check-list C) ❑
(d) For key transportation parameters
     (see Check-list D)   ❑
(e) For distribution channels and methods
     (see Check-list D)   ❑

Description of the particular risk: . . . .
. . . .
. . . .
. . . .
Description of measure and source of data: . . . .
. . . .
. . . .
. . . .

Notes:

2 Changes in market conditions

(a) Relative raw material prices
(see Check-list C) ☐
(b) Prices of competing products
(see Check-list D) ☐
(c) Trade agreements (see Check-list D) ☐
(d) Consumer tastes and expectations
(see Check-list D) ☐
(e) Currency relationships
(see Check-list D) ☐

Description of the particular risk: . . . .
. . . .
. . . .
. . . .
Description of measure and source of data: . . . .
. . . .
. . . .
. . . .

Notes:

CHECK-LIST C — PRODUCTION

IDENTIFICATION OF MODULE BOUNDARIES

1   What is the overall purpose of the investment under study?

(a) R&D project   ☐
(b) Product design project   ☐
(c) Production facility investment   ☐
(d) Marketing investment   ☐
(e) Investment in company image   ☐
(f) Other (specify)

. . . .

. . . .

. . . .

This helps identify the main investment project and the aspects of production
that are being reviewed.

Notes:

2   What is the purpose of the investment in production under study?

(a) New production plant   ☐
(b) New production technology in an old plant   ☐
(c) New waste treatment process in an old plant   ☐
(d) Improvement of existing process technology   ☐
(e) Alteration of existing production to make
     a new or changed product   ☐
(f) Improvement of present production site   ☐

The type of investment is important for other components of the appraisal. It indicates what alternatives may exist for this investment. It gives an idea of whether the investment in question is an environmental one or a general investment. If it is the latter, the analyst knows that the environmental aspects may require special care, such as an extension of the time horizon of the investment analysis, if this investment is to be compared with others.

Notes:

3   If this is an investment in a production facility, on which of the following is the main impact of the investment?

(a) R&D   ❑
(b) Product design   ❑
(c) Marketing of the product   ❑
(d) Marketability of the product   ❑
(e) Environmental performance of the company   ❑
(f) Procurement   ❑
(g) Cost accounting   ❑
(h) Product portfolio   ❑
(i) Financial position of the company   ❑

Consideration of these aspects of the investment helps to identify all the potential areas of costs and benefits arising.

Notes:

COSTS

COSTS RELATED TO PRODUCTION

1   Cost of raw materials or process materials, such as solvents, lubricants, cooling water, catalysts and so on

(a) Present market prices and quantities required   . . . .

(b) Emissions from each raw material and new
     compounds   . . . .

(c) Quantities of recyclable raw materials and the
     cost of recycling of each   . . . .

Cost of raw materials and other process materials is assessed in terms of what the company pays for these to its suppliers and *not* in terms of what they cost society or costs during the whole life cycle of the materials. Transportation cost of these materials is assumed to be part of the price the supplier charges. This includes energy used for transportation.

Emissions from each raw material may vary and may form new compounds. Either the initial emissions or new compounds may be regulated. Identify the quantities of emissions per ton of output or other suitable measure. The treatment of these will incur a cost to the company.

Quantities of recyclable materials need to be identified in order to quantify the cost of recycling. Recycling may require different technology that adds to the overall process cost and investment.

Total cost   . . . .

Notes:

2 Use of energy in production and its cost

The energy referred to here relates directly to production and not to transportation or production of raw materials or intermediates used in production, nor to transportation of end-products away from the production site. It does refer to the use of energy in transporting materials in production and for transfers within the site. It also includes the energy used in waste treatment that is an integral part of the production process. Energy used in a separate waste treatment facility within the production site is included in the overall waste treatment cost. Energy that is locked up in the material itself (the so-called feedstock energy or calorific value) is not costed here, as it does not constitute a cash outflow for the company.

Total cost . . . .

3 Cost of other materials

By-products (including energy) can be resold or reused. Are there any? Identify and quantify them. Identify the cost of reselling or reusing them. If there is no natural market for the by-products, and hence a market needs to be created, calculate the market creation costs.

Total cost . . . .

Notes:

4 Cost of waste treatment

On the basis of emission identification and quantification, make an evaluation of which emissions need to be treated either on site, at outside facilities or not at all. The cost of each alternative needs to be estimated and included in the overall cost of the investment under study.

This cost type is the one most commonly ignored in investment calculations. It must be added to analyses where a new technology or plant is under consideration. It is particularly important if there are several process and product design alternatives available to achieve the company goals set for the investment under study.

Waste treatment costs include:

(a) Cost of getting/maintaining permit for,
    and building, the facility . . . .
(b) Cost of tranporting the material to be treated,
    if not treated at the production plant . . . .
(c) Operating costs of the facility, such as energy,
    labour, maintenance, insurance, chemicals, water . . . .
(d) Training staff to use the facility . . . .
(e) Cost of monitoring the efficiency of the facility . . . .
(f) Storage and disposal of waste and residual
    products after treatment . . . .
(g) Cost of equipment and machinery . . . .
(h) Cost of selling the residual or by-products
    from waste treatment . . . .

Total cost . . . .

Notes:

5 Other operating costs of the production facility

(a) Maintenance . . . .
(b) Training of staff . . . .
(c) Cost of safety measures . . . .
(d) Insurance of the plant for hazards,
accidents and employee health . . . .
(e) Transportation required within the plant
due to production planning . . . .
(f) Storage . . . .
(g) Quality assurance cost . . . .

Total cost . . . .

6 Costs due to production location

(a) Cost of compliance due to environmental
regulation (noise or odour reduction, atmospheric
releases, ground water protection, environmental
fees and taxes) . . . .
(b) Cost of site alterations for production purposes . . . .
(c) Cost of infrastructure improvements
(water, sewage, energy, transportation) . . . .
(d) Cost of security arrangements . . . .
(e) Cost of employee transportation to site
arranged by employer . . . .
(f) Cost of relocation of staff required . . . .
(g) Cost of staff recruitment and training . . . .
(h) Cost of legal and other expert advice . . . .
(i) Local property taxes . . . .

Total cost . . . .

Notes:

7　Financial costs

(a)　Currency risk protection cost . . . .

(b)　Commodity price coverage, such as futures . . . .

(c)　Interest paid on working capital required . . . .

Total cost . . . .

8　Asset investment

(a)　Land for production site . . . .

(b)　Equipment and machinery (including vehicles,
　　　IT equipment) . . . .

(c)　Buildings . . . .

(d)　Facilities related to infrastructure
　　　(eg, power stations, waste holding tanks) . . . .

Total cost . . . .

9　Liability costs

(a*)　Future liability costs . . . .

(b)　Health insurance costs . . . .

(c)　Loss insurance cost . . . .

(d*)　Environmental damage insurance cost . . . .

Total cost . . . .

Notes:

10* Monitoring environmental costs

(a) Labour used for monitoring/reporting emissions . . . .
(b) Equipment used for monitoring emissions   . . . .
(c) Cost of outside entities used for monitoring
     emissions   . . . .

Total cost   . . . .

These costs are often ignored in investment calculations as they are usually
included in general overheads and therefore not easily identifiable.

11 Cost of regulatory compliance

This can be due to a chosen policy on behalf of the company to anticipate
future requirements. It is also possible that the process or product does not
currently achieve the required emission or other environmental limits due to
inferior or inadequate technology.

Total cost   . . . .

12 Environmental fees, taxes and required compensation

These can represent costs for discharge licences or taxes imposed for usage
of waterways, landfills and so on. The compensation may be ordered payable
to, for example, fishermen or farmers whose livelihood may suffer from the
emissions.

Total cost   . . . .

13 Direct labour cost

(a) Operating personnel   . . . .
(b) Plant maintenance personnel   . . . .
(c) Plant management   . . . .
(d) Social costs of all these personnel categories . . . .
(e) Employment taxes   . . . .

Total cost   . . . .

COSTS IN OTHER PARTS OF A COMPANY DUE TO PRODUCTION

1 Procurement costs

These are costs of purchasing raw materials, equipment and other necessary items for setting up production and operating. Other procurement-related costs — such as customs fees, VAT and other indirect taxes, licence fees, transportation costs and the like — are usually taken into account as part of the cost of materials and equipment.

Total cost   . . . .

2 Cost of accounting

(a) Quality cost monitoring   . . . .
(b) Environmental cost reporting and monitoring   . . . .
(c) Production accounting   . . . .
(d) Project accounting   . . . .
(e) Product cost accounting   . . . .

Total cost   . . . .

These costs are usually reported in overheads and not separated into categories of their own. Although this may be correct and adequate for continuous operations, it is necessary to cost these items separately when evaluating a project. A new process technology may result in complex quality monitoring and environmental reporting.

3 Cost of marketing (see Check-list D)

Total cost   . . . .

Notes:

COSTS IN INCREMENTAL CALCULATION

1 Possible costs

(a) Additional fines and fees due to environmental
    legislation if this investment is not adopted   . . . .

(b) Additional waste treatment costs if this investment
    is not adopted   . . . .

(c) Reduced sales if this investment is not adopted   . . . .

(d) Additional administrative costs if this investment
    is not adopted   . . . .

(e) De-bottlenecking production to fit this investment . . . .

(f) Cost of changes required in existing production
    facilities   . . . .

Total cost   . . . .

2 Change in production level

(a) Due to a more complex process   . . . .

(b) Due to poorer quality performance   . . . .

(c) Other reasons and respective costs of reduced
    production level   . . . .

Total cost   . . . .

3 Cost of changing procurement contracts due to changes in process technology or product design

Total cost   . . . .

4 Other cost types as listed above, but on incremental basis

Cost type   *Amount*

. . . .

. . . .

. . . .

. . . .

. . . .

REVENUES

REVENUES FROM PRODUCTION

1   Revenues from production due to this investment   . . . .

2   Revenue from reuse of raw material and other process material   . . . .

3* Revenue from sales of process by-products   . . . .

4   Revenue from sales of patents and licences   . . . .

5   Revenue from grants and subsidized credits   . . . .

6   Revenue from prizes and awards   . . . .

7* Revenue from recycling   . . . .

REVENUES IN INCREMENTAL CALCULATION

1* Savings from reduced raw material, energy and other material use   . . . .

2   Savings from operating costs   . . . .

3* Reduction in future liability costs   . . . .

4* Increase in revenues from sales of recycled materials and components   . . . .

5   Savings due to a favourable production location   . . . .

6   Savings in financing costs
(a) Change in raw material use resulting in
lower stocks   . . . .
(b) Less price change protection, such as futures,
required   . . . .
(c) Less financing needed for capital assets due to
different process technology   . . . .

7	**Savings in insurance costs**	. . . .
8	**Savings in accounting and monitoring required**	. . . .
9*	**Savings in fines for non-compliance**	. . . .
10	**Savings in taxes, fees and licence costs**	. . . .
11	**Reduction in procurement costs**	. . . .
12	**Value of increase in sales**	. . . .

Notes:

RISKS, ASSOCIATED PROBABILITIES AND DATA SOURCES

1* Changes in the regulatory environment concerning the production process

(a) Emissions to air, water, land
(b) Noise and odour emissions
(c) Safety and health aspects

Emissions to various media are already regulated for many compounds. The prediction of changes in legislation therefore involves predicting the next level of limits, the addition of new substances and the timing of such changes. This can be done on the basis of information from industry associations, lobbying groups and company specialists.

Changes in safety and health regulations can be sudden, due to a new scientific or medical discovery. It is therefore in the interests of companies and industry associations to monitor the situation carefully. Company doctors and other health and safety professionals can provide assistance. Actuaries also use statistical means for predicting various health events in populations. These statistics are available from the actuarial profession in many countries.

Employee health is of interest to insurance companies who are asked to provide health insurances for companies and their employees. Insurance companies therefore keep statistics of certain health events. Some health problems may be due to unknown effects of certain chemicals or compounds. Prediction of occurrence of these problems is difficult, but perhaps a contingency, either for total liability or the insurance, should be considered.

Description of the particular risk: . . . .
. . . .
. . . .
. . . .
Description of measure and source of data: . . . .
. . . .
. . . .
. . . .
Expected monetary value (EMV) weight: . . . .

2   Changes in regulations concerning waste treatment

(a) Particular technologies
(b) Particular emissions from waste treatment
(c) The levels of pollution remaining after treatment

The best source of information for various waste treatment technologies is the suppliers and vendors. They need to monitor the expectations of regulators in the different countries to which they are selling their technology. Engineering contractors can also provide information. Companies need to rely mostly on outside expertise and use the information obtained for selecting probabilities for likely outcomes.

Description of the particular risk: . . . .
. . . .
. . . .
. . . .
Description of measure and source of data: . . . .
. . . .
. . . .
. . . .
Expected monetary value (EMV) weight: . . . .

Notes:

3   Changes in raw material prices

(a) Changes in relative prices between alternative raw materials
(b) Changes in prices due to world/local market fluctuations
(c) Changes in prices due to relative currency values

These types of change are very difficult to predict and it is preferable to select a whole set of prices and related probabilities. Another possibility is to cover part or all of the price risk by buying futures, either in currencies or in commodities, whichever is relevant. If possible, the price variations can be transferred to selling prices. In this case the marketing department has to make an evaluation of how much price increase could be tolerated by the market.

A similar type of analysis is required for the prices of saleable by-products. These should not be forgotten, particularly if the quantities of by-products are significant as measured in monetary terms.

Description of the particular risk: . . . .
. . . .
. . . .
. . . .
Description of measure and source of data: . . . .
. . . .
. . . .
. . . .
Expected monetary value (EMV) weight: . . . .

Notes:

4* Risk of an unexpected environmental hazard

Description of the particular risk: . . . .

. . . .

. . . .

. . . .

Description of measure and source of data: . . . .

. . . .

. . . .

. . . .

Expected monetary value (EMV) weight: . . . .

Notes:

CHECK-LIST D — MARKETING

IDENTIFICATION OF MODULE BOUNDARIES

1   What is the main concern of marketing in this context?

(a)	Evaluating market prospects	☐
(b)	Studying trends in consumer tastes and expectations	☐
(c)	Studying market reactions to specific product design aspects	☐
(d)	Looking for potential markets, either product niches or new geographical areas	☐
(e)	Monitoring competition	☐
(f)	Promoting products	☐
(g)	Lobbying for the company and industry	☐

Notes:

COSTS

Marketing costs may be hidden within overheads. Although a marketing department carries out many activities on behalf of a product group or a whole company, some of them can be directly assigned to a product or investment. Marketing that focuses on environmental issues needs to be separated from the general marketing work, because it is usually directly linked with a product group or product. The cost of such marketing work, be it lobbying or special market studies, may be significant and therefore must be identified.

When marketing costs are included in an investment appraisal, it is easiest to do it in the form of a marketing budget. The budget may then be allocated between overall marketing costs and direct, product-specific marketing costs. These can be treated in different ways in the company cost accounting. In the investment appraisal phase they should be assigned specifically to a project.

Estimating sales volumes is usually based on market studies that are conducted either by the company itself or by an outside consultant. The newer and more unknown the product is to the potential market, the more likely it is that an in-depth market study is required. Often such a study is beyond the resources of the producing company and the services of an outside consultant are required.

Notes:

MARKETING COSTS IN THE INVESTMENT UNDER STUDY

1 Studying market prospects

(a) In-house market studies
(b) Externally supplied market studies

Costs should include:
* direct labour   . . . .
* indirect labour   . . . .
* materials and equipment   . . . .

These costs are often regarded as overheads and are not reported separately in the cost accounting of a company. Even if this is the case, it is important that these costs are taken into account separately and are made project-specific, as they may be significant. This is particularly true if the product is to be sold in many geographical markets. Each market will require its own study.

Total cost   . . . .

2 Studying trends in consumer tastes and expectations

These studies can be done either on a continuous basis, as a part of the overall work of the marketing department, or as separate studies for different markets and products. Again it is normal that these costs are considered as overheads, but due to the potentially large amounts that are spent on these studies, the costs should be separated out and included in the project costs.

Costs should include:
* direct labour   . . . .
* indirect labour   . . . .
* materials and equipment   . . . .
* purchases and outsourcing . . . .

Total cost   . . . .

Notes:

3 Studying market reactions to specific product design aspects

These studies can be complex and cover not only the home market but foreign markets as well. Sources of information can vary from consumer tests and test marketing to contacting potential client industries. Typically, these studies are not considered as specific projects by the marketing department and therefore it may be difficult to determine the cost of this work.

Total cost . . . .

4 Looking for potential markets

Identification of markets is one of the main tasks of marketing departments. This can be done as a continuous exercise but can also be a specific effort to create a market for a particular product. The work comprises analysis of product qualities and trying to find matching market needs. Although this work is often regarded as part of the marketing overhead, it is important at the project appraisal phase to assign a marketing budget to the project.

Total cost . . . .

5 Monitoring competition

This should be regarded as part of the specific marketing budget of the project under study.

Total cost . . . .

Notes:

6 Product promotion

(a) Advertising . . . .

(b) Trade shows . . . .

(c) Entertainment of potential clients . . . .

(d) Lobbying . . . .

(e) Other typical promotion activities . . . .

These costs are often part of the project budget if the project has a specific marketing budget. Every change to the product or production process may lead to changes in the marketing strategy. These changes have their own costs which need to be accounted for in the project appraisal.

Total cost . . . .

7 Lobbying

This is typically a task performed either by the environmental officer (manager or director) or by the company public relations department. Often this task is also shared by top management and partially delegated to outside organizations, such as trade associations and the like. The cost of this activity is seldom directly attributable to a project, but it is still important to single these costs out of the overall budget of lobbying of public relations work if the lobbying activity is directed to a particular process or product.

Total cost . . . .

Notes:

8* Eco-labelling

A consumer or intermediate product may benefit from an eco-label. In most countries this process costs, not only in terms of time and effort but also because registration may carry a fee. These fees vary greatly from country to country and from scheme to scheme. Before eco-labelling is contemplated, study the potential effect of eco-labelling on sales. Include the cost of this study in the overall marketing budget of the project.

Total cost   . . . .

MARKETING COSTS IN INCREMENTAL CALCULATION

These costs accrue only if the project under consideration is undertaken and may occur in all the categories 1 to 8. These costs are 'new' in the sense that they would not exist without this investment.

Cost type	*Amount*
. . . .	. . . .
. . . .	. . . .
. . . .	. . . .
. . . .	. . . .
. . . .	. . . .
. . . .	. . . .

Notes:

REVENUES

REVENUES DUE TO MARKETING

1   Revenue from new markets and client groups . . . .

2* Revenue from a product or production features . . . .

3   Revenue from successful promotion . . . .

4   Revenue from successful lobbying . . . .

REVENUES IN INCREMENTAL CALCULATION

1   Revenue from a price premium for better quality . . . .

**2* Revenue from a price premium for an
     environmental image** . . . .

3   Revenues from new markets and client groups . . . .

4   Revenues from increased sales . . . .

Notes:

RISKS, ASSOCIATED PROBABILITIES AND DATA SOURCES

1   Risks due to changes in market expectations

On the basis of a market study, a company can establish its expectations for market demand over time and markets. The company should investigate what the probabilities of the various market demands are at various price levels and at various points in time. These probabilities can then be included in the investment calculation.

There may be several aspects of market expectations; those of regulators, of pressure groups, of consumers and so on. Each group influences the demand for a product in different ways. These ways must be identified and the impact on demand defined.

In this way a company may end up with conditional probabilities: 'If public opinion moves to demand a total ban on CFCs, the regulator will demand a total ban in five years time with 70% certainty'. These types of dependencies need to be identified, the most important ones selected and the probabilities calculated.

Description of the particular risk: . . . .
. . . .
. . . .
. . . .
Description of measure and source of data: . . . .
. . . .
. . . .
. . . .
Expected monetary value (EMV) weight: . . . .
. . . .

Notes:

2   Risks due to changes in transportation requirements

The transportation of many process industry products is already heavily regulated, yet there may be changes and new requirements that affect the cost of transportation from the production site to the customer. Although these costs are not always borne by the producer, some may be. Again it is important to establish the most likely level of change and the probabilities of variation around it.

Description of the particular risk: . . . .

. . . .

. . . .

. . . .

Description of measure and source of data: . . . .

. . . .

. . . .

. . . .

Expected monetary value (EMV) weight: . . . .

. . . .

3   Risks due to changes in distribution channels

A product that is environmentally acceptable may require different distribution channels from a product that is less so. This may be due to wholesaler interest or different market expectations and tastes. A company may therefore have to choose different distribution channels. The cost of the most likely channel and the likelihood of being able to use it needs to be estimated. Other cost scenarios have probabilities attached to them, which are then included in the analysis.

Description of the particular risk: . . . .

. . . .

. . . .

. . . .

Description of measure and source of data: . . . .

. . . .

. . . .

. . . .

Expected monetary value (EMV) weight: . . . .

. . . .

4   Risk of changes in trade agreements

Trade agreements between countries affect the end price of a product due to customs duties and various taxes and standards that are imposed upon imported products. These may constitute a major obstacle for entering a particular market and may also change the relative market position of a product overnight. It is therefore important to monitor trade negotiations between countries and to try to estimate the impact of various scenarios on the sales of a product. Choose the most likely scenario, and assign a probability to other possible outcomes.

Description of the particular risk: . . . .
. . . .
. . . .
. . . .
Description of measure and source of data: . . . .
. . . .
. . . .
. . . .
Expected monetary value (EMV) weight: . . . .
. . . .

Notes:

5 Currency relationships

The relative strength of currencies affects both production costs and selling prices. In this way currency fluctuations present a double threat to a company. Although there are many ways of protecting a company from currently fluctuations, these may be costly, and therefore out of the question. Currency fluctuations can also be very erratic and large, which make them even more threatening to a company. A company must estimate its sales income on some level of an exchange rate. The probability of this fluctuating more than x% should be estimated and incorporated into the investment analysis.

Description of the particular risk: . . . .

. . . .

. . . .

. . . .

Description of measure and source of data: . . . .

. . . .

. . . .

. . . .

Expected monetary value (EMV) weight: . . . .

. . . .

Notes:

CHECK-LIST E — COMPANY IMAGE

Company image influences the perception of company stakeholders. These
stakeholders belong to many categories. Those of main interest here are
insurance companies, banks and other credit institutions. Their perception of
the environmental performance and image of a company affects whether
they will require a risk premium in the pricing of their services. The exact
mechanism varies from company to company and from country to country.

COST OF CREATING A COMPANY IMAGE

1   Image promotion

(a) Publications   . . . .

(b) Campaigns   . . . .

(c) Other promotional activities   . . . .

Total cost   . . . .

**2*   Activities in other parts of a company that influence the
environmental image and performance of a company**

(a) R&D   . . . .

(b) Product design   . . . .

(c) Production processes   . . . .

(d) Marketing   . . . .

Total cost   . . . .

The costs of these activities are defined in their respective check-lists.

Notes:

3   Cost of public consultation

This cost can be one-off for a specific project. It can also be a part of the overall operating cost. It covers such areas as public health studies, building community partnerships, applying for licence to operate and negotiating with local authorities. The time and cost involved should not be underestimated, as Test Case B shows (see page 88).

Total cost

4   Cost of poor image

Most companies are concerned about their public image, particularly if their products are closely associated with the company. This has led many companies to adopt various corporate programmes, such as Responsible Care and publishing environmental reports. Yet the impact of these efforts on company image is not necessarily positive. Unexpected events may actually lead to a deterioration of company image. Before taking any improvement actions, companies should carefully evaluate the impact of poor image on sales and other activities.

Total cost

Notes:

REVENUES

**1   Increased sales of a product, product group
or all products** . . . .

2   Reduction in financing cost . . . .

3   Reduction in insurance premiums . . . .

All these forms of income take time to accrue and it may be difficult to time the revenues. It may also be difficult to allocate the revenue to a specific product or to trace it back to a specific activity undertaken by a company. Such incomes may also be difficult to single out from other revenues. A company must discuss the effects of a good image with its various stakeholders in order to identify the present status and be able to monitor changes.

Notes:

RISKS, ASSOCIATED PROBABILITIES AND DATA SOURCES

The main risk associated with a company image is that the image depends on human perception. As discussed in Chapter 3 and Appendix 1, there is no reliable measure for company environmental performance. Through discussions with credit institutions and insurers, a company can gain an understanding of how the various environmentally-geared improvements may alter its credit rating and financing costs or its insurance premiums. Changes in other stakeholders' perceptions can depend on events that may be outside the control of a company — such as major accidents in other companies in the same industry. These types of uncertainties and related risks are part of the overall business risk of a company.

Description of the particular risk: . . . .

. . . .

. . . .

. . . .

Description of measure and source of data: . . . .

. . . .

. . . .

. . . .

Expected monetary value (EMV) weight: . . . .

. . . .

Notes:

APPENDIX 2 — EXPECTED MONETARY VALUE, OR EMV

A2.1   RISK ASSESSMENT

Traditional investment analysis techniques (for example, net present value (NPV) and internal rate of return (IRR)) are 'no-risk' measures of the time value of money. They assume the projected (annual) net cash flows will occur with certainty, both in terms of their timing and amount.

At the time of analysis exact values for all the factors which will affect the investment cannot be measured — for example, future exchange rates, inflation, production process efficiency and so on. However, one can estimate ranges of possible values that each variable may have.

There are various options to allow for risk in assessing an investment:

(1) Sensitivity analysis — this provides a result for a given change, but it does not consider the likelihood of this change occurring.

(2) Descriptive language — 'high', 'low' risk, 'a long shot', etc. This is a purely subjective view, since the definition of risk, and levels, differs from person to person. The interpretation of this language also differs from person to person.

(3) Use a higher discounting rate in discounted cash flow (DCF) calculations. This, however, is not appropriate. The discount rate used in evaluating future cash flows should relate to the earnings rate of future invested capital at the *company* risk level (which is not the risk level of an individual project); the discount factor serves only to tie together the value of money at two different points in time.

(4) Quantify the relative likelihood of occurrence of each possible outcome — that is, risk analysis. The financial value of each outcome can then be weighted by its likelihood of occurrence, and a 'risk-weighted average value' of a project/investment calculated, taking account of all the possible outcomes. This risk-weighted average value is called the expected monetary value (EMV).

In order to derive the EMV consider:

- the possible values for the variable parameters in a scenario;
- their relative likelihood of occurrence (expressed as probabilities summing to one);
- the risk-weighted average value for the parameter derived from this probability series.

The EMV of a range of possible financial values for a parameter is the risk-weighted average of the possible values associated with its probability. To derive the EMV for a total project, sum the EMVs of the individual elements within it.

The scenarios under consideration differ not as scenarios with differing levels of each parameter but as strategic solutions to a problem in a company. Often sensitivity analysis is conducted by varying the expected value of a parameter included in the equation. In the case of EMV scenarios, the scenarios are different in terms of the actual parameters included and the solution they reflect. The two EMV scenarios can be an end-of-pipe solution and an intrinsically clean production technology to achieve a certain emission level. The traditional approach to scenario analysis would have been to study different configurations of the end-of-pipe technology.

Like the NPV of a project, the EMV or a sum of separate EMVs is not found in company finances later. Both depend on the actual outcome of the cash flows and the actual discount rate of the company. These in turn depend on the operating environment and the financial markets.

A2.2   CALCULATING EMVs

The EMV approach requires the specification of three possible values for each parameter (as shown in Figure A2.1), with the probability of each given at predefined levels:

- the *most likely* value;
- the value which there is a 90% probability of exceeding as the *lower* value;
- the value which there is only a 10% probability of exceeding as the *upper* value.

With these three values it is possible to calculate a factor by which to move the *most likely* value to the *risk-weighted average* (mean) value — that is, the EMV for financial values.

The area of the triangle L, VL, PI in Figure A2.1 is 0.1 (if the question is framed to give the '90% probability' value). So is the area of the triangle U, VU, P3. The area of the triangle L, P2, U is equal to 1, since this is the sum of all the probabilities. From this five statements can be made:

164

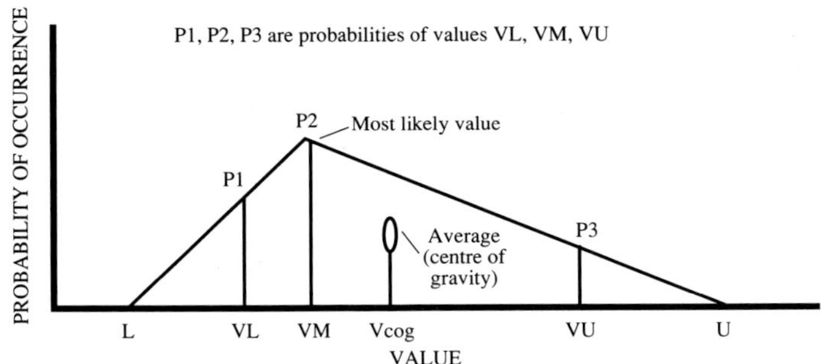

Figure A2.1 Calculating EMVs.

VL = *lowest* value, with P1 chance of this being realized;

VM = *most likely* value, with P2 chance of this being realized;

VU = *upper* value, with P3 chance of this being realized;

L   = the lowest value, with zero probability;

U   = the highest value, with zero probability.

$$(VL - L) / 2 \times P1 = 0.1$$

$$(U - VU) / 2 \times P3 = 0.1$$

$$(U - L) / 2 \times P2 = 1$$

$$P2 / P1 = (VM - L) / (VL - L)$$

$$P2 / P3 = (U - VM) / (U - VU)$$

where values for VL, VM and VU are known. Solution of these equations gives values for P1, P2, P3, L and U. With the actual parameters of the triangle known, the position of the 'centre of gravity' can be calculated; this is the weighted average point in the probability distribution, and its position on the value axis represents the EMV of the values given for the parameter under inspection. These calculations can be done using a PC spreadsheet application.

A2.3   STEPS IN DETERMINING EMVs FOR PROJECT PARAMETERS

The following practical steps can be followed when compiling the relevant EMVs of the various parameters in the overall investment evaluation.

(1) Determine the parameter. In most cases this is clear. Sometimes a surrogate measure is needed. This is particularly the case when future liabilities or company image need to be evaluated.

(2) Estimate the annual cash flows for each year within the time-scale. The cash flow of a certain parameter may vary from year to year — for example, the amount needed to invest in capital assets. Calculate the gross cash flow for each year for the parameter. This is easiest done on a spreadsheet where every parameter is listed in the first column. The following columns represent each year in the calculation. In the matrix that is thus formed fill in the relevant gross cash flow. See the example in Test Case A, page 63.

(3) Determine the risk factor for each parameter. At this stage ask the questions outlined in the previous sections of yourself or the person with the best knowledge of the risk profile of the parameter in question. Do ask these questions exactly as given. Otherwise the questions may be stated wrongly and give the wrong answer for the risk factor.

(4) Multiply each parameter value in each year with the risk factor. This is also most easily done on a spreadsheet containing the non-risk-weighted values of each parameter.

(5) Calculate the net risk-weighted cash flow for each year.

(6) Discount the net risk-weighted cash flows of each year to the base year, using the company discount rate.

GLOSSARY

Accounting profit

In the Western world accounting is usually done on a so-called accruals basis. This means that income and expenses are matched in an annual profit calculation according to the period they refer to, but not according to the period in which the actual cash flow has taken place. Therefore a sale may be included in turnover although money has not been received from the customer. An electricity bill can be included in the costs although the bill has not yet been paid.

Benefit

Benefit to a company is usually measured in monetary terms. A benefit to a company is a positive impact on the company profit and cash position — that is, cash inflow.

Depreciation

In accounting its annual result, a company is allowed to include only a fraction of the cost of an investment into the total expenses of the company that particular year. This fraction is called depreciation. An asset that the investment has created can be depreciated over the expected lifetime of the asset. In some countries there are accounting laws or customs that regulate over which time the asset is to be depreciated.

Direct costs

Costs that relate directly to a product or process. Typically, these are labour and material costs that can be found in the production area. In investment analysis this definition has to be enlarged to include the costs in marketing and other non-production departments that relate directly to the investment under consideration.

Discounting factor

Discounting is a way of evaluating the future cash flows of an investment in today's money. Thus discounting is actually compounding in reverse. In compounding, interest is accrued on interest and capital; in discounting, the operation is reversed. The rate at which the reversing is done is called the discount rate or discounting factor.

Discount rate
The rate at which discounting is taking place should reflect the company's rate of gain from alternative investments. This depends in turn on the rate at which the company can borrow money, and hence it reflects the risk the company is perceived to entail by the capital markets. In order to determine their discount rate, companies need to calculate their weighted average cost of capital (WACC).

Eco-balance
These balances try to reflect the way a company affects nature by using energy and raw materials and, in some cases, also other natural resources, such as clean air, scenery or clean water. There is no one definition for an eco-balance.

Eco-label
An eco-label is a label (usually a logo) that a producer can try to obtain for a product to signal to consumers how environmentally friendly the product is. The schemes are usually administered by local boards representing industry, government, environmental groups and consumers. The conditions for qualifying for a eco-label are often based on the life cycle effects of the product.

Expected monetary value (EMV)
The expected monetary value of a parameter in an investment calculation is the discounted, risk-weighted value of a cash flow in the future.

Expenditure
In accounting legislation there is a differentiation between the way in which investments and expenditures are to be treated in the annual profit and loss account. Expenditures are items that have to be fully deducted from the year's income; investments are depreciated (that is, only a fraction of the total cost is deducted from the year's income).

Externality
A spill-over effect or a third party effect that has been caused involuntarily is called an externality. In this book, it is an item that is outside the system — the company. The financial impact of an externality may need to be brought into the evaluation of an investment if it is expected that the externality may cause a cash inflow or outflow to the company at some stage.

Fixed costs
Costs that do not vary in the short term with the variation of production levels

are called fixed costs. This definition of costs is valid only if the production conditions are relatively stable. It can normally be discarded in investment evaluations because the time horizon is long.

Hurdle rate

In order to be able to choose among various investment options, companies set themselves various decision rules. These often come in the form of hurdle rates. An investment needs to yield a net present value (NPV) of a certain size to overcome the hurdle and be accepted. Usually the NPV is to be expressed as a percentage over the investment required to achieve the NPV.

Indirect costs

Costs that relate only indirectly to a product or process are indirect costs. Typically, these include items outside production. In investment evaluations these costs relate to general company administration, accounting and finance, and to legal costs.

Internal rate of return (IRR)

IRR is the discount rate at which an investment has a zero net present value. The decision rule is to accept an investment for which the IRR is higher than the company's WACC or the interest the company would earn if it invested the same amount elsewhere.

Investment (see Expenditure)

Liability costs

Liability costs are those that a company may have to carry if it is found liable for causing a hazard. Often a surrogate measure needs to be found for these costs.

Life cycle analysis (LCA)

In life cycle analysis (LCA) the impact of a product on nature is studied from the moment raw materials are extracted for the production of the product, to the moment the product is disposed of. The analysis covers areas such as raw material consumption, energy use in production and transport, emissions during the production and use, and during disposal.

Net present value (NPV)

NPV is the value of discounted future cash flows today.

Opportunity cost

In discounting terms, opportunity cost is the cost that is accrued when money is not invested and thus interest is lost. In general terms, it is the potential income that would be lost if the intended investment were not made in the first place.

Out-of-pocket expenses

These are expenses that a company has to pay in real cash. Some costs are not cash flows in accounting a company's annual profit but can nonetheless be regarded as costs. In investment calculations real cash flows are the most important ones. Non-cash items are usually needed only to calculate the accounting outcome of an item, such as taxes.

Overheads

Overheads are costs that companies accrue due to various administrative tasks or for tasks that have not been divided into each individual department or product. An example of the first is personnel management and accounting; of the second the use of a machine, such as a copying machine or a drilling machine.

Payback period

The time over which the investment is recouped is the payback period. If the investment was £2000 and the income from the investment is £1000 p.a., the payback period is two years.

Profit

In accounting terms, profit is the positive difference between a year's income and expenses. Profit can also be measured for other units than the whole company — for example, a product or a department.

Profit sustenance

Profit sustenance means efforts for keeping profits at a stable level from year to year.

Return on investment (ROI)

ROI is accounting-based income as a proportion of net book value of an asset. Net book value is the value of an asset that remains on the company books after depreciation is made for the year.

Revenue

This can mean income, earnings and profit. In America, it is used to mean sales.

Risk

Risk is the uncertainty around the outcome of a parameter. It is usually measured by a probability of an outcome of a parameter.

Sunk cost

A cost that has already been accrued in the past does not enter the calculation of future costs and the net present value of an investment. In this sense, it is a sunk cost.

Total cost assessment

This methodology, developed by the Tellus Institute in the US, presents an enlarged set of costs and benefits and a longer time horizon to be included in the investment evaluation of environmental investments. Risk is incorporated by using an appropriate discount rate.

Variable costs

Costs that vary in the short term with the variations of production levels are called variable costs. They are not to be mixed up with direct costs that are costs that relate directly to a product or process.

REFERENCES

1.	Tellus Institute, November 1991, *Alternative Approaches to the Financial Evaluation of Industrial Pollution Prevention Investments*, 55.

2.	Umweltbundesamt: Berichte 11/91 *Umweltorientierte Unternehmensführung*, 495.

3.	White, A. and Becker, M., 1991, *Total Cost Assessment: Catalyzing Corporate Self-Interest in Pollution Prevention*.

4.	Tellus Institute, November 1991, *Alternative Approaches to the Financial Evaluation of Industrial Pollution Prevention Investments*.

5.	Umweltbundesamt: Berichte 11/91 *Umweltorientierte Unternehmensführung*.

6.	White, A. and Becker, M., 1991, *Total Cost Assessment: Catalyzing Corporate Self-Interest in Pollution Prevention*, 329.

7.	Tellus Institute, December 1991, *Total Cost Assessment: Accelerating Industrial Pollution Prevention through Innovative Project Financing Analysis*, with applications to the pulp and paper industry, 3–6.

8.	Meffert, H., Benkenstein, M., Schubert, F. and Walther, Th., 1986, (8a), *Unternehmensverhalten und Umweltschutz, Arbeitspapier Nr.31 der Wissenschaftlichen Gesellschaft für Marketing und Unterrehmensführung, Münster 1986*, 20.

9.	Umweltbundesamt: Berichte 11/91 *Umweltorientierte Unternehmensführung*, 558.

10.	Meffert, H., Benkenstein, M., Schubert, F. and Walther, Th. (8a), 1986, *Unternehmensverhalten und Umweltschutz, Arbeitspapier Nr.31 der Wissenschaftlichen Gesellschaft für Marketing und Unterrehmensführung, Münster 1986*.

11.	MacLean, R.W., 1987, *Estimating future liability costs for waste management options, GE Corporate Environmental Protection*, November 1987.

12.	Brealey, R.A. and Myers, S.C., 1988, *Principles of Corporate Finance*, 3rd edition (McGraw-Hill Book Company).

13.	American Institute for Pollution Prevention and EPA, *A Primer for Financial Analysis of Pollution*.

14.	*ENDS Report* (12a) No. 215, April 1993, p.25.

15.	Norsk Hydro, September 1992, *PVC and the Environment*.

16.	Ryding, S.–O. and Steen, B., July 1991, *EPS-Systemet, B–1022* (Swedish Environmental Research Institute).

17.	Mäkelä, U., Supponen, M., February 1993, *Ympäristömerkintä* (Eco-labelling) (Tekes Teollisuussihteeriraportti).

18. Shigeyuki Hashizume, *Environmental Labelling in Japan*, The Eco Mark (Japan Environment Association).

19. Wicke, Haasis, Schafhausen and Schultz, 1992, *Betriebliche Umweltökonomi* (Verlag Vahlen, München), 542/3.

20. Interviews with Association of British Insurers: Mr Gerry Rutlidge, 10 August 1993; Mr Mark Watthorn, St. Paul International Insurance Company; Mr Jeffrey Pickering, Zurich International (UK) Ltd, 11 August 1993.

21. Gordon, T.A. and Westendorf, R., 1989, Liability coverage for toxic tort, hazardous waste disposal and other pollution exposures, *Idaho Law Review*, 25: 567–618.

22. Pruett, M.C., 1990, Environmental cleaning costs and insurance: seeking a solution, *Georgia Law Review*, 24: 705–732.

23. Krozer, J., June 1992, *Decision Model for Environmental Strategies of Corporations* (Institute for Applied Environmental Economics. The Netherlands).

24. Reinikainen, T., January 1993, Suomen suurimmat pistekuormittajat vuosina 1990 ja 1991. The largest point emission companies in Finland in the years 1990 and 1991, *Ypäristökatsaus* (The Finnish Water and Environmental Agency).